OCEAN BOUND
WOMEN

Sisters Sailing Around the World in the 1880s

The Adventures • The Ship • The People

OCEAN BOUND WOMEN

Sisters Sailing Around the World in the 1880s

The Adventures • The Ship • The People

ANDERS HALLENGREN
Stockholm University, Sweden

World Scientific

NEW JERSEY • LONDON • SINGAPORE • BEIJING • SHANGHAI • HONG KONG • TAIPEI • CHENNAI • TOKYO

Published by

World Scientific Publishing Europe Ltd.
57 Shelton Street, Covent Garden, London WC2H 9HE
Head office: 5 Toh Tuck Link, Singapore 596224
USA office: 27 Warren Street, Suite 401-402, Hackensack, NJ 07601

Library of Congress Cataloging-in-Publication Data
Names: Hallengren, Anders, 1950– author.
Title: Ocean bound women : sisters sailing around the world in the 1880s—
 the adventures—the ship—the people / Anders Hallengren, Stockholm University, Sweden.
Description: New Jersey : World Scientific, 2023. | Includes bibliographical references and index.
Identifiers: LCCN 2021057719 | ISBN 9781800610897 (hardcover) |
 ISBN 9781800610965 (paperback) | ISBN 9781800610903 (ebook) |
 ISBN 9781800610910 (ebook other)
Subjects: LCSH: Women and the sea--Sweden--History--19th century. |
 Sailing--History--19th century. | Atlantic (Sailing ship) | Sailing ships--Sweden. |
 Seafaring life--History--19th century. | Voyages and travel--History--19th century. |
 Söderström, Maria Mathilda Charlotta, 1867-1947. | Söderström, Emmy Axelina Charlotta,
 1868–1944. | Sisters--Family relationships.
Classification: LCC G540 .H29 2023 | DDC 910.4/509252 [B]--dc23/eng/20211216
LC record available at https://lccn.loc.gov/2021057719

British Library Cataloguing-in-Publication Data
A catalogue record for this book is available from the British Library.

Copyright © 2023 by World Scientific Publishing Europe Ltd.

All rights reserved. This book, or parts thereof, may not be reproduced in any form or by any means, electronic or mechanical, including photocopying, recording or any information storage and retrieval system now known or to be invented, without written permission from the Publisher.

For photocopying of material in this volume, please pay a copying fee through the Copyright Clearance Center, Inc., 222 Rosewood Drive, Danvers, MA 01923, USA. In this case permission to photocopy is not required from the publisher.

For any available supplementary material, please visit
https://www.worldscientific.com/worldscibooks/10.1142/Q0321#t=suppl

Typeset by Diacritech Technologies Pvt. Ltd.
Chennai - 600106, India

'What we are? and Whither we tend? We do not wish to be deceived. Here we drift, like white sail across the wild ocean, now bright on the wave, now darkling in the trough of the sea;—but from what port did we sail? Who knows? Or to what port are we bound? Who knows? There is no one to tell us but such poor weather-tossed mariners as ourselves, whom we speak as we pass, or who have hoisted some signal, or floated to us some letter in a bottle from far.'

'Introductory Lecture on the Times', R.W. Emerson (1803–1882)

'"As o'er the past my memory strays," I recall certain reminiscences of a past generation ... which may possess some interest to readers ... and, therefore, I will record them, without delay, as my own generation is nearly past, and the one prior to that, entirely gone, and soon there will be no one either to tell or to listen with interest.'

'Some Reminiscences of the Warminster Society', Mary W. Early (1914)

Contents

Preface .... *xi*
Acknowledgements .... *xv*

   I. Aye Aye, Destiny! 1
      1. Ashes to Ashes 1
      2. Seafarers 3
      3. The Life Journey 5

  II. Maids, Orphans and Women's Rights 9
      1. Background 9
      2. The Plague 11
      3. The Workhouse Rules 12
      4. At the Orphanage of Salem 14
      5. There Was No Time for a Proposal 18
      6. The Return of the Grim Reaper 22
      7. The Importance of Learning 29
      8. The Great Liberation 32

 III. The Travel Diary 39
      1. From the Gulf of Bothnia to Øresund and Zealand 41
      2. The North Sea 44
      3. Under the Tropic of Cancer 47
      4. Celebrating Christmas on the *Atlantic* 49
      5. The Line-Crossing Rite of Passage 53
      6. Life in the Blue 56
      7. The Albatross 59
      8. To Australia: *Adelaide* 61
      9. *Melbourne* 69
    10. Ball with the Imperial Russian Navy 77

11. Crossing the Pacific Ocean 80
12. '—It sometimes seems to me that I would like to rebel against myself, against God and the whole world!—' 83
13. Pitcairn and the *HMS Bounty* Mutineers........................84
14. The Stars are Out: Under the Southern Cross 89
15. 'Love without Expectation' 92
16. Mid-Ocean in the Intertropical Convergence Zone 92
17. Port Townsend, USA 96
18. Among Rothschilds and American Indians 99
19. Port Gamble—A Meeting Place for Travellers From Afar...101
20. Forest Walks 104
21. Little Boston and Day Cruises 108
22. Return to Port Townsend...111
23. On the Sea..114
24. Back to Australia—Crucial Days 116
25. South Melbourne Vistas and Manners 119
26. The Monster Wave that Struck the *Orient*....................121
27. Christmas in Australian Summer 124
28. '*Frankly, nowadays I do no longer dare to be sincere and honest with myself!*'...124
29. Interlude: The Dream of True Love and the Perfection of Matrimony 125
30. Romance: Recollections of Love and Longing— Heading for Micronesia 136
31. Arrival to Malden Island..142
32. Remembrances of Malden Island 147
33. Out Fishing with the Islanders......................................150
34. Alteration of Course to South-East 10,000 km 152
35. Storm Damage—With a Leak Towards Cape Horn 154
36. Passage of the Horn at the 56th Parallel South...............158
37. *Portland ho!*...160

IV. The Voyage in Context and Retrospect 163
   1. Origins 163
   2. Trade Routes...165
   3. Prevailing Sailing Conditions..168

4. The Gefle Harbour 168
5. The Master Mariner 171
6. Loadstars 172
7. Astronomy and Meteorology 174
8. Sea Rhythms 176
9. Life on the Sea 178
10. Authority and Management 181
11. Embarking the *Atlantic* 183
12. Biography of the *Atlantic*—The Early Years 187
13. The First Series of Voyages 190
14. Setting Sail 192
15. The Take-off in 1885 193
16. Close-up Picture of the Indispensable Chief Officer 195
17. The Living Conditions of the People 196
18. Contemporary Celebrities and Events 199
19. Emmy's Recollections 203

V. The Last Days of the *Atlantic* 211
1. Nuptials 211
2. Mate's Memories 215
3. From Argentina to the United States 220
4. Steering for Australia and the Cape Horn 223
5. Life on Land 226
6. Time Lapse: Long Voyages in a Changing World 228
7. Söderström's Long Farewell Voyage 229
8. Murder Mystery in Cape Town 231
9. The Last Days of the Skipper King 236
10. How Does This End? 239
11. Load Line and Waterline 245
12. The Appearance of the Longshorewomen 248
13. Perseverance to the Utmost Limit 251
14. Fighting to the Bitter End 253
15. The Wreck 254

VI. What Happened Next? 255
1. Wartime 255
2. Ghosts 261
3. Family Affairs 267

VII. Voyage to Unexplored Regions of the Earth.........................277
  1. Family Business 277
  2. A Coincidental Encounter 281

Bibliography: Sources and Literature...*283*
List of Illustrations...*297*
Index *301*
About the Author...*317*

Preface

THINGS CARRY TRACES of those who once owned them. The passed-on person's pockets are emptied of contents, and the contour of a personality is sensed. In the trunk of heirlooms, fleeting life stories are glimpsed, lives without equivalent. What do you choose to take with you when you have to leave everything? What does the fugitive bring; what passes from one life to another and retains its meaning? In every human being's life, there are stages of crisis and breakup, parting and arrival without return. The one who feels reborn and finds a new haven or heaven, does not leave everything behind. In what one carries is found a mirror of identity. In its light, we, survivors, can vaguely see a person's image.

As heir, in this case of a huge collection of documents, photographs, and memorabilia stored in two seaman's chests, one in metal and another of camphor wood, together with comprehensive genealogical and research material, collected and saved by descendants and passed on to me, I find myself as the mouthpiece and spokesperson on earth of the dead and dumb on the other side of the grave—in the long chain uniting us in the arc from ashes to ashes, all holding on to the same anchor cable.

The renowned author Irmelin Sandman Lilius, born in 1936, who has written similar biographical documentaries and travel books, is calling me up from Finland exhorting me: 'You have now the sole responsibility for making these people seen and heard when you are still among the living, and no one else can accomplish this.' Next, the American poet and novelist Dale M. Kushner sends me a letter from Wisconsin impelling its addressee: 'You certainly have inherited amazing and astonishing tales that are the makings of a Scheherazade experience of fabulous adventures that leave the reader enchanted and wanting more. I would have to say the Fates gave you this task of bringing your ancestors into a new dimension and into the hearts and minds of others. This feels like a somewhat holy mission, which I know sounds a bit highfalutin' and inflated, but nonetheless, the diaries and materials have been given to you at a time when you are able to absorb them and write.'

These are urgent requests, as oppressive as elevating in a way, summoning me to get all this done, and the story now to be told for the first time is extraordinary, and happens to be of universal interest. Thus, the ancestors, who are hailing from the oblivion of forgotten traces and artefacts materialize, become present in their absence, and the author serves the present and posterity by honouring the departed—*they* providing us with unique experiences and knowledge of the unwritten past, while rising from the shadows again to assume the unstinting role as givers, not takers.

The following is the history of the Swedish sailing ship *Atlantic* (1876–1911) and its crew, including its master mariner Johan Axel Söderström and his seafaring family of Gefle,[1] which travelled to all parts of the world as part of their international shipping trade activities. A main feature of the book is the first diary from a sailing trip round the world written by a North European woman, and one of the first ever, which stands out as a uniting text for the years 1885–87 and connects to all earlier and later events in the storyline. The sources for this chronicle consist of manuscripts, documents and accounts collected from the descendants along with oral traditions and personal memories; all hitherto unpublished and unknown.

The protagonists are two motherless girls who, after foster home and a pioneering girl's school, went to sea, took on a sailing ship and

---

[1] This east coast seaport at the mouth of the *Gavle River*, the stream which debouches into the southern part of the Gulf of Bothnia in the Baltic Sea, has passed through various naming customs through the centuries. In the nineteenth century, and in the beginning of the twentieth century, it was called *Gefle*, which is used here, and the etymology of the ancient name is still a matter of dispute. Reminding of the Swedish *gavel* ('gable'), it may be distantly related to the German *Gabel* ('fork') or the Greek κεφαλή ('head'), being originally a topographic appellative. In 1940 it was decided that the right spelling of the place name henceforth would be *Gävle*.

travelled the oceans for years, trips with amazing consequences: Maria 'Mia' Mathilda Charlotta Söderström (1867–1947)—seafarer, diarist, and her sister Emmy 'Emy' Axelina Charlotta Söderström (1868–1944)—seafarer, pianist, entrepreneur, and storyteller.[2]

So, the special feature of the book is Mia's continuous journal, *Anteckningar under sjöresan* ('*Notes during the sea voyage*') of 7 November 1885–25 September 1887, handwritten in Swedish in an extant memorandum-book in the family collections, translated and annotated in this book and carefully put in context (Figures 7 and 8, Ch. III).

Forerunners among female circumnavigators are not many. They were, first of all, Madame Rose de Freycinet's famous French *Journal* of 1817–21, in an edited translation known as *A Woman of Courage*, followed by the remarkable Austrian explorer and naturalist Ida Laura Pfeiffer, who made the trip twice between 1846 and 1855, and was preceded by the scientific accounts of a fellow-countrywoman, the pioneering amateur botanist Jeanne Baret's travel around the world in 1766–1770.

The main document was handed down to the present author along with notebooks, records, ship lists, remembrances and various *anas*; narratives, annals, and other material from forebears—a long line of shipmasters, seafarers and shippers (both patrilineal and matrilateral). In the cornucopia of reports there is an account from Pitcairn Island about descendants of the Bounty mutiny. Meetings with indigenous people. The narrative encompasses lively meetings with local residents and Native Americans in the Northwestern USA, and various encounters with Polynesians and other peoples of Oceania. It includes a unique account of a prolonged stay at the rookery and guano fields of Malden Island in the mid-Pacific Ocean, also named Independence Island, which was later laid waste by British nuclear tests.

In other respects of special interest, is a ball with the Russian navy, and the vivid depiction of a season-length stay in Australia, where one of the sisters in the event married the captain of another ship, who was later killed in Cape Town.

---

[2] As there is more than one speaker in this book and the perspectives shift between family context and scholarly comments, both the pet names—which they preferred—and the registered first names will occur as identifiers in this book.

We also get access to the adventures of forebears and descendants, who kept up the seafaring traditions, and are enabled to see the voyage in context. And moreover to learn about how two of their children became shipping agents, and another one who as mariner got involved in the First World War campaigns. Evidence is also given for the role in voyages of discovery, including the strange story on how an island in the Southern Ocean was named after one of their descendants.

The author has collected material about all people mentioned in this book, and done extensive archival research, but principally has amassed recollections and papers from old relatives. He has also travelled in the wake of the ship *Atlantic* and paid visits to a number of places where these people have been—some houses where they had decisive experiences still standing and partly preserved—at times going by sea, on rare occasions also navigating in some of those distant waters.

In particular, the author's first-hand acquaintance with Oceania was an asset to the writing, but his knowledge of barques is mainly academic. In real practice, that is limited to a day aboard the Russian tall ship *Kruzenshtern*, still in operation, when he had the opportunity to study the splendour of its riggings. ∆

Acknowledgements

This book could not have been written without the help, learning, and enthusiastic attention of Mia's grandchildren, the late Yngve Herrström, and Stig Herrström, Chinon, in the first place, who both urged me to write this book and provided me with source material; her granddaughter Ingrid Gurli Maria Stedman; her great-grandchildren Thorleif, Christina, Pontus Carle and Madlen Herrström, the last two in France; Emy's children Sven and Gunborg, and her granddaughter Siv Pestliki in Greece; her great-granddaughter Maria L. Hallengren whose knowledge, research, and collections were indispensable; her grandsons Björn Hallengren, the travelling sales representative Roland Bergendal, and Yngve Bergendal, the builder and navigator of the ketch *Mickan* (1973) who carried a picture of captain Axel Söderström in the cabin—and finally went to Cape Horn, followed the traces of Nordenskjöld in the Southern Ocean, and travelled the transatlantic trade wind route with the barque *T/S Gunilla* and wrote about it; Maritha Karling, the daughter of Emmy Söderström's son, the shipping agent Bo Hallengren—all genealogists and collectors of family memories and memorabilia. Furthermore, the research that has been done by the eminent maritime historians Ingvar Henricson (d. 2015) and Sven-Olof Nordenberg, who in different ways kindly and substantially assisted me in my investigations.

I also owe a lot to Admiral Stefan Engdahl, of the Royal Academy of Naval Science, for taking an interest in my work, of decisive importance, along with the invaluable intellectual support of professor Neil Kent at Scott Polar Research Institute at Cambridge University, and Alexander Malmaeus, chairman of the Anglo-Swedish Society in London, of which the present author is a life member. I am also grateful to the sympathetic and well-versed WSPC acquisition editor Stephen Soehnlen in Warsaw for his surprise call during the pandemic, enthusiastically asking me for a new book, which in the event made the World Scientific Europe imprint in London commission this work; in due consequence also many thanks to the efficient production editor Michael Beale who became my first guide at the Covent Garden office.

The obliging and well-informed support of a number of institutions and specialists has also been indispensable: Melissa Rydquist and Cecilia Marnetoft of the National Maritime Museum in Stockholm; the Swedish National Archives—with great praise to royal archivist Mats Hemström; Landsarkivet in Härnösand; Länsmuseet Gävleborg headed by the eminent archivists Sofia Hedén and Anki Börjesson; Arkiv Gävleborg; Ann-Kristin Karlsson and Johanna Holmkvist at Gävle Kommunarkiv; the archivist Åsa Lundborg of the Stockholm City Archives; Marianne Larsson, Helena Lindroth, and Leif Wallin of the Dept. of Cultural History at Nordiska Museet in Stockholm; the cynologists and ethologists Lotta Bengtsson and Staffan Thorman; image editor Johanna Nilsson of Fotoskolan, Stockholm, who carefully restored nineteenth-century family photographs, as did the graphic designer, Paolo Sangregorio; the Royal Library in Stockholm; the Royal Museums Greenwich, the British Library and the Natural History Museum in London.

I am also under obligation, for their helpfulness, advice, hospitality or resourcefulness, to the Returned and Services League of Australia (RSL) and in particular John Jones at the New South Wales Branch; the staff of the Princess' Theatre, and of the Museums Victoria—including the Immigration Museum and the Royal Exhibition Building; the Ibis Hotel Melbourne, pastor Katja Lin at the Church of Sweden in Sydney, now in Melbourne; Jean-Claude Herremans FRZS and my former colleagues at the Linnean Society of NSW at Botany Bay; my landlords G. Moreau in Brighton-Le-Sands and O. Heldon of the Hurstville Society in Penshurst; the laureate poet John Tranter in Balmain, NSW; dir. Neville Jarvis; the National Library of Australia in Canberra, Victoria, the dean and staff of Christchurch Cathedral in New Zealand, and Karen Wright at Waitaha; the University of Hawai`i at Manoa, Oahu; the Good Hope Centre in Cape Town, the Colegio Universitario San Gerónimo de La Habana; Dr Pedro Pablo Rodríguez at the

Centro de Estudios Martianos; princess Winchinchala of Wampanoag; MP Amineh Kakabaveh; the late Sir Derek Walcott of Saint Lucia; Asociación de Estudios del Caribe; Jeff Kildahl, Dept. of Social and Health Services in Tacoma, WA; and the always accommodating alma mater Harvard University in Cambridge, Massachusetts, USA, where I was once taught the art of scholarly writing by my supervisors Daniel Aaron and Harry Levin, both born in 1912.

As a writer I also owe a debt of gratitude to professor Elena M. Chekalina, Moscow, and in particular the novelists Dale M. Kushner in Madison, Wisconsin, and Irmelin Sandman Lilius in Hanko, Finland, for prompting me with benevolent stage whispers. A tribute of respect, finally, to my late friends and longstanding sources of inspiration and encouragement, professors Johny K. Johansson in New York and Vera L. Gancheva in Sofia, both deceased in 2020. ∆

I. Aye Aye, Destiny!

1. Ashes to Ashes

It all started with the city conflagration of July 1869 when the two surviving siblings Mia and Emy in frail and vulnerable toddlerhood spent the night outdoors at the House of Nobility Street in the East District. They were sheltered by a fortepiano covered by a large blanket, as all houses burned down and the coastal town Gefle was ablaze in one of the largest fires in Swedish history. The 47-year-old glazier and councillor Jacob Hallengren, a practical man who in his childhood had seen a Brownie doing good deeds, organized rescue operations for the sufferers.

Close to North Sea Customs Street in the same district, quarters of old populated by seamen and fishermen, the sisters for a while were to dwell with their ailing mother in temporary lodgings. It was a homely shack in an area walled by a decayed fence of planks, where the rats swarmed the huge rubbish-heap in the enclosure. The youngest boys, too small for fishing trips in the vast harbour, used to slink into that forbidden and forbidding area, and, sitting at the fence, angle for fat hauls with their long rods.

The garden was the playground, and one of the boys remembered how he happened to hit a patrolling policeman's helmet with a football, and the ensuing moment of fright when the constable stopped, found the doorway to the property, and in a majestic manner stepped into the grounds with the leather ball under his arm and the imposing sabre at his belt.

When Emy's and Mia's mother died at 24 in 1873—whose mother had died at 27 in the cholera epidemic of 1853 leaving the daughter to be adopted at an orphanage—their father was at sea. Their longing for him, which was always hard to bear, became increasingly unendurable.

By then their father was in Jamaica with the full-rigged ship *Thor* and returned home the following year—and only then found that he had become a widower and the children motherless. The postal service, by mail boats, was much worse than today, and sometimes slower than the merchantmen. Deceased crewmen, not taken by the high seas or buried at sea, accidentally returned home in coffins or caskets with transport vessels to shocked mothers and widows without prior warning.

Not until 1895—the year when Emy gave birth to her oldest son, the father of this book's author—Marconi invented the wireless telegraph. The first Atlantic cable was laid in 1858, but telegraphists were never found on the boats these ancestors carried and rarely in the ports they called at.

The technology was limited to sextant, compass, nautical charts, log, lead plummet, barometer, thermometer, marine chronometer, wind meter, ship clock, hourglass, drawing set in brass with a pair of compasses, draft pin, protractor and transversal scale, as well as field glasses. All indispensable on astronomical navigation out in the oceans with no land in sight, and especially at night, gearing up and steering with the help of the navigation stars and the nautical almanac, which indicated coordinates, declination and hourly angle for fifty-eight stars, the vessel doing well over five knots per hour on the average and a hundred nautical miles round the clock.

Emy remembered that early one morning she woke up hearing, as in a dream, an anchor chain go in a windlass at a berth in the harbour a mile away and immediately discerned that it was daddy coming home at long last. She referred to this occurrence in her early childhood as a clairvoyance of love.

At the funeral in the small wooden chapel, the girls had been sitting in their loneliness silently watching their mother, who was lying pallid on *lit de parade* displayed in an open boat-shaped coffin, surrounded by solitary strangers, and in particular they noticed an old, veiled mourning woman sobbing, whose identity they never learned. People kept at a distance; there was a fear for contagion and the dead in the phthisis plague that claimed so many victims.

2. Seafarers

Captain Johan Axel Söderström, known as the Skipper King, grim and determined, who used to command his men in storm as in calm, sat pale with them at the sepulchre chapel after his arrival to the home-port. His weather-beaten face, a picture of a thunderstorm that could shed light in the impenetrable darkness at sea, was like an empty slate where the text had just been erased leaving a thin film of moisture or, expressive but speechless, an immobile rock hit by the hurricane. He was known for finding the right course of the sailing ships by the winds that caressed his cheeks.

With his large hands hanging heavy at the sides, his gaze was rigidly turned towards a remote distance, riveted to an invisible horizon. The ringing of the church bells on Sunday waxed and waned in the wind of time, moved back and forth, was coming and going. 'Goodbye, she had said. God bless you. Father will take care of you.' Withering away and emaciated by consumption, on the iron bed of sooth in the waiting room of death, her soul finally had departed before their eyes.

The daughters sat silent, looking down at the floor. They held each other's hands. They had not yet begun school and were still at kinder-garten. What would be their fate now? What dangers did not lurk on land as at sea?

> Full is the earth of evil,
> and full of it is also
> the immeasurable sea

the ancient sermon echoed. 'May Providence guide us on our journey!' The ship must soon let go and set out for the world. Was there any greater honour for the family than setting a full-rigged ship to sail for new goals? What other options were there, what other way out? The crew waited, new rents had to be taken, shipping bargains to be struck. We must submit to our destiny and put up with it. What else could be in store for us? The ocean is our home as much as the cottage in the eastern quarter were. And what is a home that you cannot build and carry inwards?

The captain had made it all clear and stood resolute. 'In the future I want to show them the world as I saw it in my absence far away during all their childhood, while the old water-wheel of the idyll turned in the Gavle river and vagrant cattle on the road frightened the infants and reluctant skippers kept announcing "Remain off-shore in rough seaway." Yet may God preserve the man who comes too close to them! Good they will be in the father's supervision for years onwards.'

However, only 5 and 6 years old, the daughters were left ashore in care of the orphanage of their mother, the private children's home Salem and the girls' school for ten years, until the time was ripe to sign on a ship, whilst their father without postponement left the port for the oceans again and stayed away for ages.

In 1876 he became master mariner of the recently built barque *Atlantic*, a grand three-masted vessel, which would determine the fate of his family and not least the life of his daughters, who took on in September of 1885.

From Emy and other people in the crew stem a number of stories about the people and adventures at sea, in addition to strange experiences at distant places abroad. Notable is the journal Mia kept on board in 1885–1887, which includes the first account of a circumnavigation written by a Scandinavian woman—the last in the Age of Sail, and one of the first ever in women's literature, preceded by Ida Pfeiffer's travels in the 1850s, and the Frenchwoman Madame Rose de Freycinet's pioneering letters of her romantic seafaring adventures on board the corvette *Uranie* of 1817–1820.

The reasons why there were just a pocketful of such diaries preserved worldwide were by no means any lack of ability to write or any paucity of woman authors or female journal writers—in the nineteenth century most educated women kept diaries and wrote letters—but the scarceness of girls travelling round the earth in a circular fashion, and the fact that woman seamen had been considered an anomaly if not a paradox. Sailors in many countries thought that women on board were calamitous, and the great explorers of the past did not bring any in their crews. The covenant on the sea was that between the male crew and the ship of female gender forming a union, a steadfast pact bridling the forces of nature for better or for worse.

Over and above this rationale there are very concrete and plain explanations too. 'World tourism' and 'globetrotting' are fairly newfangled things, which are hardly older than the very words used to denote these modern phenomena (late nineteenth century). Furthermore, there is a business-related cause to consider. In fact, for economic reasons sailing trips round the world were still uncommon in the early 1800s, and often unpremeditated. These circular tours were more of interest to explorers, adventure lovers and research expeditions than to commercial travellers. Indeed, the Magellan–Elcano expedition accomplished the feat as early as 1519–1522, but the first circumnavigation of a Swedish ship was that of the brigantine *Mary Ann* in 1839–1841, with Captain Nils Werngren at the helm of this two-masted vessel. In the merchant fleet, that craft was a famous predecessor of the magnificent *Atlantic* of 1,000 metric tons, carvel-built of Swedish oak and pine wood and coppered, which sailed on until 1911.

3. The Life Journey

Mia thought that she would vex posterity with her conundrums on the clandestine and never-ending journey. Once, at an imperial navy ball, she had found an elegant cavalier, a stylish officer of the Russian armada, forlornly stupid. He had not even heard of her homeport Gefle right across the same narrow Baltic Sea—a sister city in the waterways of the era, which in the middle of the nineteenth century had the largest merchant fleet and next to Gothenburg was the major harbour in her country of origin.

Her nieces Gunborg and Greta later recalled, that one day in their childhood they were told that twenty-three of their seafaring kinsfolks, men and women, were currently on board or at the quay somewhere, and the number of captains, skippers, mates, coxswains and other seamen in the line of ancestors forty-five.

After all she knew that she was the first North European woman ever to write a journal of travel on a circumnavigation, spending motherless her teenage years with her sister among men on a slow deep-sea craft, a merchantman with pending destinations and unknown destiny, ocean bound year after year like her ancestry for eons, the voyage an inward journey carrying her fate, longing and fear.

She would never have dreamt of or imagined the range of adventures and tragedies that were to follow. That the sisters were to be acquainted with so many outlandish distant places and end up in Australia, or that her only son would become a war victim in the Gallipoli Campaign. Nor that they would, on their long journey, encounter aboriginal populations in the Pacific including one or two earnest cannibals, besides privateers and mutineers, and pass through scores of ordeals, and face strange, cruel fatalities worldwide—storms, calamities and shipwreck—and experience long dark nights in despair, lurching from one crisis to another as the boat did.

It was a time of adventurers and the moon still held mysteries. It was possible to hear how the ghosts of the past weave through our days, stir the heart and make the aloneness less lonely. The sea was near, and gulls, fulmars, huge albatrosses with wingspans of 10 feet, shearwaters, petrels and frigate birds traversed the oceans as kindred spirits, rapturous in the wind. Whereas navigation was astronomic, fate was astrological.

Haunting sailing crafts and steamers in the fog, nightmares, mirages and illusions were countered by prayers and hymns in the scream of gale-force winds, and when the hurricane struck at its worst the order was *All hands on deck!* which also meant *Up in the lookout and into the rig* to trim the sails and save the ship, *Anchors out!*

At times, *Man overboard!* was as common as *Land ahoy!* The captain was granted with almighty authority as chief, police, judge, punisher, doctor, priest and was often forced to combine these roles, being the only one carrying firearm *en route*, a Webley & Son .442 five-round black powder revolver with 11, 1-mm lead bullets, the ship equipped with old-fashioned swivel guns mounted at gunwales and the master bestowed with the right to conduct funerals at sea. Attacks and losses were common.

Still, this generation's Holy Grail was the dream of a Camelot that would never end. The latter half of the nineteenth century was also an era of mass migration, which was not seasonal as in the animal kingdom of their wandering companions travelling the air and the waters along with them, but with no return, on occasion without hope, too, ending in *cul-de-sacs*.

These seafarers believed in the *anima mundi*, the world soul, and wondered what that might be trying to tell us about our unknown fates if we could listen. In her diary at sea, the ship tossed by the roaring ocean, the diarist noted down in despair: 'Will I ever get married?'

On 10 April 1995, the present writer received a moving and unexpected answer to her lasting query in the physical form of a family heirloom. I was then to an old pewter maker named Hopp at 2 Banner Street in Stockholm, a lone remnant of an old Central European guild of skilled craftsmen. The large inherited tin dish in my possession he identified as a Carolina plate from 1762. According to an inscription, which he could decipher, it turned out to be a wedding gift to Maria Söderström ('MSM') in 1888, the writer of the diary.

I was told that the pewter dishes from the 1700s at that time had become a common gift from the Age of Freedom and its bygone days since their uses as dining services were completely driven out of competition by porcelain in the 1800s. They had become increasingly rare and precious collector's items and primarily for show as ornaments.

The tin founder further observed that this dish had spent a large part of its life, about a hundred years, at room temperature indoors, but not in the middle of a table but standing on edge at a wall as a decoration. This he perceived from its oval shape. 'Pure tin is soft metal,' he explained, 'and is slowly shifting shape in the course of times by the force of the earth's gravitation, as is also taking place at a much slower rate with much harder materials such as glass and rocks.' Wash the dish with detergent and soft brush under running water, he recommended, and 'use it every day as a bowl for colourful fruits like lemons or oranges, it's beautiful!' In a flash of memory, I then recalled that at the barque *Atlantic*, one citrus fruit a day was compulsory for the crew to prevent scurvy, both under the command of Axel and his successor, his son-in-law. (Curiously enough, by the force of the long-accustomed habit, descendants have stuck to Captain Söderström's ordination to this day.)

So, one day she would get married, under remarkable circumstances, as it turned out, on this long and adventurous voyage, despite her fear and misgivings of never having a home, and a family of her own. And the sisters' descendants were to hand down her journal and all notes and stories and family archives to me for research and relation.

As will transpire from the hitherto unpublished documents in the following, all translated from the Swedish handwriting, she had married the captain of another ship in Melbourne, and in course of time her father, the sea captain, sank to another sea while her husband, who thus succeeded her father as commander of the *Atlantic*, met his death under mysterious circumstances in Cape Town, South Africa.

An island of the Antarctic in the Southern Ocean was to be named after one of their descendants.

II. Maids, Orphans and Women's Rights

1. Background

THE GRANDMOTHER AND GREAT-GRANDMOTHER of the sisters Emmy and Maria Söderström were poor maids with several fatherless children. Their mother was a charity girl from an orphanage.

Before the invention of household appliances, washing machines, refrigerators, vacuum cleaners, electricity, telephone, automobiles, tin goods, semi-manufactured and sanitary articles, disposables, nurseries, nursery schools, day-care centres, and especially before the introductions of birth control and public elementary school, maids were an inalienable part of Swedish households. Most of them arrived from the countryside in their early teens to become servant girls in town for a year or two, but also to learn cooking, domestic work and baby care for the future marriage—and relieving their own families from maintenance obligations.

For lack of alternatives, some stayed for life in this occupation—a subordinate, exposed and vulnerable position where the probability of harassment and abuse to some point increased with age. When the state of dependence had become solidly established, maids were often treated at their employers' discretion. To those unmarried maids who became pregnant and gave birth to illegitimate children, there were not many options left, and the dependence was unconditional. In particular, to be a breadwinner and support a family as a single parent was hard, as the maid's salary or rather compensation was usually just food and lodging. So much the worse for them, they were invariably condemned as bad women by the bigoted paragons of virtue.

In late eighteenth-century Sweden, about 5% of all children were born out of wedlock, but there were considerable dissimilarity between the social classes, and also between different parts of the country. The proportion of 'illegitimate' children was greater up here in Norrland than down in southern Sweden. The church banned sex before marriage and was repudiating unmarried women who gave birth to children. However, the priests were more severe in the south and more forgiving and tolerant up in the northern and more poor parts of the country. In Gefle in Gästrikland, there were rarely explicit critical notes in the house interrogation lists and the parish registers about unmarried women with children. In Lillhärdal in Härjedalen, moreover, two out of five women who were brides had already given birth to a child, with no questions asked. The proportion of illegitimate children then increased until about the middle of the nineteenth century, when our story begins.

Emmy's and Maria's great-grandmother, Anna Greta Olofsdotter, was born in 1798 in the parish of Österlövsta, Upland. She went the 31 miles northwards through Skutskär to Gefle in her fair youth to serve as a maid with the mayor, among others. In her late 20s, she considers that she has had enough of city life and moves home to Åkerby in Österlövsta with her then 1-year-old daughter Carolina Petronella Löfström, born in 1825. Greta will get three more daughters, too; father unknown.

Her daughter Carolina leaves home in Åkerby in October 1844 to seek her fortune as a maid in Gefle as well. Like her mother, she became pregnant and gave birth to her son Carl Johan in July 1846, and later on also Amalia Mathilda on 14 April 1848. In April 1850, her son Per August was born, who died in July as a baby of three months. In all these cases, there is no acknowledged paternity.

From the parish registers, the whereabouts of Carolina's little family hereafter can be followed year by year. Enough to say, they moved each year. She stayed in various cottages and farmhouses with other families in Öster—'the East' in the seaport Gefle conurbation, the Fourth Quarter of the town where many fishermen and seamen had lived since old times. The Söderströms, Clases, and Zellingers and many other seafaring families had their domiciles there, as did many poor people, until this part of the city went up in flames in 1869.

Carolina lodges with a boatswain Zacharias Löfstedt; a sea customs officer L. Östrand who is a relative of the Söderströms; the skipper Engelbert Andersson; the skipper Fredrik Lovall's widow Helena; Anna, the widow of the strong iron carrier Per Österlund; and so forth. In the early autumn of 1853, the Löfström family's home is farm No. 143 in Gefle City's Second District. Many people are living in that farmhouse! It was located in the southwest corner of Västra Nygatan and Stapeltorgsgatan, just a block from Stapeltorget, which in turn was adjacent to the churchyard.

Amalia Mathilda (called Thilda), 5, at this time has no other sibling than her big brother Carl Johan, 7. Their mother is 27.

Then the plague of 1853 strikes.

2. The Plague

A cholera epidemic was approaching from the south. An infectious, highly dangerous intestinal disorder caused by bacteria—microorganisms that were not yet known, nor their cure. The very word 'antibiotics' did not even exist. Some verdant city parks of today are cholera cemeteries from the 1850s, mass graves. The course of the disease was often short. After prolonged attacks of nausea, diarrhoea and vomiting, the afflicted died from dehydration. In addition, people were afraid of the sick and avoided them—and even more the dying and the dead.

When reports reached Gefle in the summer that the epidemic had found a hold in Stockholm, the public health committee organized countermeasures to guard the citizens from the fearful and little-known enemy. All forces were mobilized and they prepared for the worst, forestalling the catastrophe by taking advance action against the contagion. The Casern was made ready to serve as emergency hospital. All town gates and customs stations were closed. Voluntary guards at the entrances relieved each other every sixth hour, working on four shifts 24/7.

The sentries were provided with lanterns and candles by the town council, as well as long tongs for safe handling of documents and new arrivals at their outposts, touching them only with bargepole.

No one knew that the pest was waterborne and spread with drinking water, waterways and wet provisions, or that it loved rains, floods and well-washed perishables, and that it struck where sanitary conditions, drainage and water were inferior, and malnutrition common—as among the poor. The East, Öster, was such a district and was affected more than other parts of the town when the infection passed the gates. Street children were provided with shoes, and foresighted persons wore copper plates on their breasts or used garlic garlands as protection against the evil.

Nonetheless, despite heroic achievements of the medical staff—including extra reinforcements of physicians, nurses, and assistants in shift work—the epidemic took its toll during August, September and October of 1853. Of a population of almost 10,000, 426 fell ill, 182 men and 244 women, most of them adults or of middle age. 141 were hospitalized, and 239 died. The corpses were covered with unslaked lime and their coffins were nailed-up and in haste buried in a specially arranged graveyard [Ljungström, 2007]. Not only poor people perished in the pestilent disease, so did also the successful naval architect and shipowner Lars Bång (b.1806), who was living next to his wharf, the northern shipyard, in the Third Quarter of the city, very close to the eastern district.

In number 143 in the Second Quarter, on the Friday of September 30, the 5-year-old Thilda and the 7-year-old Carl see their emaciated mother Carolina pass away before their eyes, guarded by neighbours keeping the distance. A female friend and colleague of their mother, and about the same age, is also there: Johanna Lindberg, born in 1826. She is living in the same house and is probably the one that is closest to the children at this moment, besides their old country-dwelling grandmother who is not present in this scene. Ms Lindberg would not forget this day and in secrecy always kept a watching eye on her friends' little motherless daughter.

The funeral took place on Saturday, on the first of October, less than a day afterwards. There was no obituary notice.

3. The Workhouse Rules

Initially, neighbours and relations did their best to take care of the young orphans, but few had the means or the situation to support and parent them. The commitment of public authorities was active and well-meaning but, in this respect, limited to poor relief. There was no public childcare system or any child welfare. Into the bargain, the number of poor and parentless was growing large.

There were alternatives, though, including an institution. There was a workhouse, which was a poorhouse for the able-bodied. The orphaned could also be placed elsewhere, as in foster homes, or; first of all: *be boarded out*.

To begin with, a solution was found for Carl, the 7-year-old boy. He first came to sea captain Johan Erik Forslund's wife Brita Catharina Dahlman in farm No. I 168, with the address Nedre Bergsgatan 10. She had been married to the skipper who in 1830–1832 sailed the brig *Idogheten*, 'Industry', in external trade. Captain Forslund had learnt to helmsman in 1827 for the renowned navigator captain Gustaf Zellinger (b. 1794), the father of the neighbourly captain Pehr Zellinger in the Fourth Quarter. Carl Johan could not stay with Brita Caisa for long, unfortunately, but sometime next year in 1854, he was sent to the Workhouse—the last resort.

This was a large three-story building in Workhouse Street, an overcrowded place with a harsh regime, and across the gate at the street one could read its serious rule:

He who does not want to work, he should not eat either.

At the Workhouse, everyone was served Rumford soup. In the middle of the century, this nutrition was still the standard in Gevalian poor care. The famous physicist Sir Benjamin Thompson, known for thermodynamics, alias Count Rumford, had introduced this economy soup for the poor in Bavaria half a century ago, along with the cultivation of potatoes. It was an early effort in cheap, scientific nutrition for the masses, and his soup was to appear in various shapes and recipes all over Europe. In Gefle at this time, it was a brown viscous decoction, consisting basically of pearl barley, peas, potatoes, herring, bread, water, salt, allspice and vinegar.

Sweden had introduced a new poor care ordinance in 1847, a framework legislation for parishes and cities, but two years earlier Gefle had adopted a special regulation for poor relief. Then, a main principle and aim was that children should, if possible, be boarded out in rural areas. That was the leading idea: outsourcing children to the countryside 'at a reasonable price with decent foster parents'. This proved difficult and risky, however; an abortive experiment with the youngest that in a handful cases came to a sad end, and in the 1860s, the regulations were changed and the principles cancelled, and consequently other kinds of accommodation called for.

When Carl Johan arrived to the Workhouse, the place was supervised by inspector Johanna Margareta Nyström (b. 1812). His stay was short, however. He is recorded for a house interrogation on 30 October, 1854, but then he must already have been consigned to a farmer in rural Valbo, to be raised as an unpaid farmhand. It is written in the note: 'boarded out to Wahlbo'.

Anyone who survived such a childhood should have been able to endure the hard life for settlers in America—which became the boy's pipe dream. Carl Johan Löfström Hammar moves to Stockholm when he is 16 years old and then continues to Sundsvall and returns to Gefle in 1864. When he has turned 20, he leaves Gefle again for Stockholm, determined to emigrate, and is thereafter no more heard of by family and friends [Herrström, 1973]. Hopefully he made it, but he is not in the Statue of Liberty–Ellis Island registry of sea arrivals.

4. At the Orphanage of Salem

But who will comfort little Thilda, then?

There was a blessing in disguise. When the plague ravaged the society in the autumn of 1853—and father, mother or both were carried off from children in many homes—many compassionate residents became involved and engaged in this concern, and in this predicament many of the elder realized the need of an orphanage.

There had been one long before. The town had founded a children's home already in 1777 at South King Street, which was closed down ten years later, replaced by the inveterate system of auctioning off the miserable kids to the lowest bid. In the despair of social breakdown, idealist citizens wondered how to invest in this project without money. Where to get funds from?

According to legend, a tiny 12-year-old disabled girl was listening in her corner of the sofa when the elderly discussed this. She was a sensitive child, and when she heard about the fatherless little children she became deeply affected. One day she offered all her savings as a donation and first contribution to the home. The story, which has been told by all who had anything to do with this institution, does not identify the girl or the sum involved, but it is known that she died shortly afterwards and that her touching act of empathy started the whole thing.

Her generosity became the trigger for residents to become interested in the orphanage, and they sacrificed willingly. A fund-raising collection started according to the principle that 'many a nickel makes a mickle'. Already in the autumn, the first step was taken towards a final solution of the problem, and shortly 'the first two orphaned boys, at the age of three, were accommodated with Mrs E. Ch. Winroth who had his home in the former Varvsgatan No. 2. With all care and love she took charge of the little lads'. The funding and the number of children grew, a large house was built for the purpose and finally, 'On 1 October 1859 Mrs Winroth could transfer the children to the new home at Islandsholmen' [Ekman, 1937, 7, 11].

What is of our principal concern here is that this childcare pioneer, who was the matron and manageress of the future Salem orphanage during its inception, also took little Thilda under her wing. Amalia Mathilda Löfström became her first charity girl. She was so dear to her protector that she was adopted. Her family name was changed to Winroth.

In short, this meant that she was saved, and in benevolent hands. Maybe there was a family connection, and that blood is thicker than water. The grandmother of her female patron Elisabeth Charlotta Leufvenius (b. 1810), Brita Elisabeth, was surnamed Hammar, married to the skipper Anders Osenius; at her christening in 1848, Mathilda had a godmother called Ebba Augusta Hammar; her own brother for some reason carried the second name Hammar.

More important, the loving mother deputy Charlotta had a nice residence in the First Quarter where she lived with her husband Olof Winroth, the navigator of a number of sailing ships—the *Audacia, Hazard, Minerva, Ocean, Oden,* and the *Sophie.* The special circumstances of the adoption can be made out from the estate inventory

Fig. 1. *Amalia Mathilda Winroth, née Löfström. Portrait photograph 1866.* Private possession. The Hallengren Archives.

drawn up on 17 October 1855—'for Captain Olof Winroth who died *without issue* during a foreign sea voyage on the 29th of June last'. Thus, it is officially stated here that he has no family, and that their marriage was childless. Something else can also be concluded from the document. Mathilda is climbing the social ladder to the middle class.

The image of a snug and well-supplied flat develops. The extraordinary long list of personal property includes: a pocket watch of gold; a ring with a precious stone; four dozen wine glasses of cut glass; tableware with twelve dozen plates; a tea set; escritoire, aquavit casket and bed with curtain, all in mahogany; a divan table, various books, a game table, meerschaum pipes, plate in silver, a gilded wall pendulum, wall mirrors with gold frames; a desk; painted blinds; an octant, sextant, marine charts, umbrellas; proceeds from the sale of the schooner *Hazard*, etc., and a balance of three thousand riksdaler, a considerable sum—the head of a ministry, a government top post, had an annual income of seven thousand riksdaler banco.

Charlotta did quite well as a widow as she also had a pension from the company of the late shipowner Bång, the one who died in the epidemic. The boys and the girl in her care were supported by the childcare society since she actually *was* the orphanage proper. Hence, to begin with, three children were accommodated on behalf of the orphanage management.

Charlotta Winroth's childcare centre was in fisherman Erik Gustaf Söderberg's old cottage on Islandsholmen, No. 47 in Gefle City's First District (I 47). She and her husband Olof Winroth had lived there since at least 1851. Olof's mother Margaretha Selgmar also had lived there, although she died on 12 November 1852. Even after her husband's tragic death at Elsinore on his way home with the schooner *Sophie*, things were going on just as usual, as she had a task to fulfil and was busy with three children.

The house interrogation record of 1856–1860 gives the children Amalia Mathilda Winroth (b. 1848), Pehr Johan Öberg (b. 1854) and Carl Olof Åsbrink (b. 1856), living on the property with this address. But Mathilda had evidently arrived earlier than the two boys, as she is already included in the house interrogation ledger of 1851–1855 for

No. I 47. It states that she is 'a foster daughter, has previously been called Löfström, and that her mother is the maid Carolina Petronella Löfström'. As we already know, Mathilda came from the big farmhouse No. 143 in the Second Quarter (II 143), where she lived with her brother and mother and Johanna Lindberg in the autumn of 1853.

When the newly built house of the Salem Orphanage is opened, only Mathilda stays in Mrs Winroth's care for the 1860s, as is evident from the house interrogation list for 1861–1865. Then the boys are no longer with them. In 1862, the number of children at the institution has grown to fifteen.

As Charlotta is still employed by this private children's home, an advantage through the years is that her adopted child has the opportunity from time to time to attend the school at the Salem orphanage. At that pioneering school, reading and writing, figures and general subjects are taught with a pedagogy based on the Lancastrian method of mutual instruction, where the pupils teach one another in small groups. Several well-informed and prominent woman teachers contributed to the remarkable success of this schooling as leaders, among them the head Sophia Grape, who succeeded Mrs Winroth in 1859; and later on, Ms Hulda Terserus.

Mathilda also attended one of the small private girls' schools, which provided personal tuition—woman teachers was the fastest growing occupational group. Among other things, she learned to play the piano during these formative years, and it is told that she had a musical ear and played well.

5. There Was No Time for a Proposal

Years pass. Axel Söderström, born 1841, a young sailor of a family of seafarers, lives in farm No. 126 in the Fourth Quarter in the East with his parents until their death in 1860 and 1862, respectively. He is the youngest of a large group of siblings where four have already passed away: Johan, Carl Gustaf, Carolina, Reinhold.

He is now closest to his two older sisters—Johanna, married to sea captain Anders Leonard Clase, and Charlotta, married to sea captain Pehr Zellinger. When Pehr dies like Napoleon at Saint Helena in 1864 and Charlotta becomes widow, Axel moves to her in Zellinger's farm in the Fourth Quarter at Riddarhusgatan, the village street fashionably named House of Nobility Street, even though there was never such a house in the town.[1]

Time to marry, settle down and form a family of his own?

After many years of apprenticeship at sea, including employments on the barque *Christina* and the full-rigger *Andrea*, and after two years of training at the Navigation School, he has become commander of the full-rigged ship *Gustaf Wasa* in 1865, just 23 years old. During the few visits to town during his years on sea, the sailor had cast his eyes ashore upon the beautiful girls sauntering in the alleys at home, and stunningly flocking on the church green on Sunday afternoon without shawls, gathering from all the parish.

He became apt to visit the amiable widow Mrs Winroth every now and then, a distant relative who had been married to a respected colleague in the seafaring world,[2] yet another man who had tragically been lost at sea. The visits were probably more frequent and longer because the foster girl Amalia Mathilda lived in her home (Figure 1). Their conversations became more and more intimate, and most of the time they sat alone together on the sofa and chatted.

One day they were taken unawares by Mrs Winroth entering the room, and who cheerfully observed how much the young couple enjoyed each other's company, and she then clapped her hands and exclaimed:

'Mina kära barn, det var ju roligt!' (*'My dear children, this was delightful!'*)

Axel later on told his relatives that he never got the opportunity to propose. After this joyful event, the matter was settled and understood. Everything went so fast, and there was no way out of it! The two were married without the least delay when he was on leave on 25 June 1866,

---

[1] A captain's mansion was often a real farmstead with domestic animals or livestock, and single-family houses have always been uncommon in Sweden.
[2] There were many connections. In 1838, Axel's father Reinhold had succeeded Olof Winroth as master of the *Minerva*.

and the wedding thus took place in Gefle on Monday after Midsummer, between Johan Axel, 24, and Amalia Mathilda, 18.

The happy event took place in the beginning of the years of climate change and famine. This meant cold summers and bad harvests, poor fish catches, ice on the Baltic Sea. Even lack of crispbread, butter, herring and dill potatoes—the staple food in town that was food for festive occasions among the porridge-eating rustics.

The week before Midsummer this year was unusually cold and windy, yet it was a little better during the weekend itself. *Gefle-Posten* of 27 June summed up the amusements of the moveable feast: 'Midsummer Day has, with the exception of a few fist fights and drinking offenses, passed peacefully. The Carousel and the dance pavilion Walhalla, the steamers with their pleasure trips, and the dram tables on the out-door-restaurants, have all been well-attended.'

The merry-go-round referred to, might have been the one at Västertull, not far from the farm No. II 143 where Mathilda lived with her brother and mother in the early 1850s, remember? That roundabout was named 'the pigswill mill' in popular parlance, as the town's maids liked to swing around there. They paid for the tour with the money they received when they sold household waste to the swineherds, the hogwash that fed the municipal porkers. This money was allowed by master and mistress to keep as a kind of *perk*—their only sort of cash payment.

The following year, in 1867, the Midsummer was really rough. It was the year when the grain did not ripen in Norrland; one of the worst years of starvation. In *Gefle Dagblad* on 11 August 1951, there was a graphic account of the years of distress in 1866–1867—when the potatoes finally were as small as peas, and the summer came so late that on Midsummer Eve people with carts drove home firewood from the forest over the ice on the Baltic Sea, and on the second day of Midsummer, they could travel by sled on the lakes.

Maria 'Mia' Söderström's granddaughter Ingrid Gurli Stedman, née Herrström, was one day in the future to pay attention to the fine old embroidered bed linen in her heirlooms, a set which she previously thought were her grandmother's and had not been touched for a hundred years. They were all kept in a cupboard. When she took a closer look at them, she saw that they are marked JAS and AMW and dated

25 June 1866. That was Axel's and Mathilda's wedding day, as appears from the town's wedding book. Surely the couple had the sheets with them on the ship *Gustaf Wasa* when the marriage bed was launched!

This is what happened. Axel succeeded his brother-in-law Anders Clase as commander of the *Gustaf Wasa* in 1865, went out in the late summer and made a short trip to the Mediterranean Sea: Gefle–Cardiff–Alicante–Torrevieja–Gefle, and returned on 30 May 1866, with salt from Spain. Subsequently, Axel signed and enrolled a new crew just over a month later, that is, in early July of 1866, and made a trip down to the Mediterranean again with timber and returned to Gefle on 9 November of the same year.

As a matter of fact, without any delay the hasty wedding had been celebrated with a very long honeymoon at sea, the husband's place of work! It went forward like a dance. The couple's daughter *Maria* was born on 16 June 1867—which shows that Mathilda was aboard on that four-month voyage with the Wasa ship and the first daughter conceived *en route*![3]

Probably Mathilda had also been with him on the journey in 1865–1866. The daughter *Emmy* was born in 1868—and she was also born in seaways and tossed on the waves. There was no end to it. Maria's and Emmy's mother Mathilda had started a new tradition: *women on the sea*, casting in their lots with the seamen.

In the nineteenth century, *Mathilda* was a very popular name in this municipality, but strangely enough, apart from a 15-m yacht, only one single Gefle ship has been named *Mathilda*: O. A. Brodin's barque from Emmy's year of birth, 1868.

---

[3] Some wives of master mariners did not reach harbour before delivery, and as a result gave birth to children in the cabin. Among others, this happened to Andrietta, the esteemed consort of Captain Fredrik Miltopaeus. The labour and the ensuing obstetrics took place on the maiden transatlantic voyage of the *Gustaf Wasa* in 1839. With the sailor Robert Rettig acting as midwife, the son Hjalmar was born on the inward passage to New York. It went off all right. The baby Hjalmar Miltopaeus grew to manhood and became auditor of the Royal Swedish Customs. Among Captain Miltopaeus successors on the *Gustaf Wasa*—including the relatives Anders Clase and Axel Söderström—the latter, in a way, kept up the production by successfully bringing the marriage bed to his stateroom on the ship.

Nonetheless, with the two toddlers, mother Mathilda had to stay home with her sister-in-law Charlotta Zellinger for a while after the seagoing adventures, to take good care of her progeny. Consequently, alone with them, she experienced the catastrophe of 1869—the conflagration that made two-thirds of the town population homeless. Eight thousand people escaped to barns, shackles, sheds, cabins, and other temporary lodgings when the smoke cleared.

Behold the disadvantage of houses made of wood! All Swedish cities and shanty towns burned down sooner or later! This happened everywhere. No more than an igniting spark in the sawdust set it all off beyond control, and all citizens got on their feet. City fires are a long story in the history of the cold and densely forested North, where stone houses were not the first choice.

A neighbour born in 1863, and some years older than Mathilda's vulnerable protégées, remembered the fatal day of 10 July 1869, in a series of images that had branded themselves on his memory. He did not recall the smell of burning, but how houses in the middle of the dry summer were ablaze, and the sound of crackling, the heat and the wind; flames of fire, sparkles and soot-flakes; how people ran about 'like ants in an ant-heap which a boy had stirred with his cane'; the fall of burning beams and the terrible roar far beyond the deserted houses; how the streets were empty; and that the walkways were crowded with people going away; others busy saving personal property and household goods before it was too late [Söderström, 1940].

6. The Return of the Grim Reaper

Aunt Charlotta Zellinger's house at Riddarhusgatan 11 was deserted after a number of belongings and furniture had been rescued and secured, and she fled along with her children Evelina, Pehr Johan, Knut and Ida Zellinger—the latter born in 1861, and a lifelong female friend of Emmy and Maria, who were saved along with their mother.

Long afterwards, upon returning to shore, Mathilda's husband was shocked to find the quarters a sooty heap of ruins and his family scattered fugitives (Figure 2).

All the East District was ravaged. His wife and children moved about between a number of temporary lodgings: they stayed for a

II. Maids, Orphans and Women's Rights 23

Fig. 2. *Emmy and Maria as toddlers, rescued from the city fire of 1869.* The Maria L. Hallengren Collection.

while with shopkeeper J. P. Öster's widow at Islandsholmen, farmhouse No. 41 in the First Quarter; barracks built or put in order for the homeless; skipper Säfström's cottage at Islandsholmen south of the Gavle River, opposite to the later erected Seamen's Church; and a year thereafter the adjacent farm downstream.

Personal property was dispersed, but much that was lost was soon replaced, and in a couple of years, many houses were rebuilt. Emmy's and Maria's father in the autumn of 1869 signed off the *Gustaf Wasa* to look after his family, see to their needs and to support them. He bought them a number of things and paid for their living, but as head of the household, and in the beginning of a great career as shipmaster, he was anxious soon to return to his profitable work—and to his home: the captain's bridge in the seaward wind. In 1870, he signed on the full-rigged ship *Thor* and waved the three goodbyes. It was a deep-sea vessel bound for the oceans, serving as the family livelihood—until the ship went down in a storm off the Azores in 1874.

His sister Charlotta, his children's dear aunt, had to put up with a good deal—the heavy blows struck one after another. Early left a widow, she had lost a 3-year-old daughter, her mother and father, and at 35 her home and neighbourhood was devastated in the conflagration. Next year, in 1870, a new family residence had to be furnished in the Fifth Quarter, with limited means.

In 1873, she had the most traumatic experience a mother could ever have, a kind of incident, by no means unique but abysmally shocking, which the improved communications of a later era were to spare families and dear ones. A ghastly and macabre event—happening, incidentally, in the gruesome reflections of later crime writers, as for example in Conan Doyle's 'The Adventure of the Cardboard Box'.[4]

Her son Pehr Johan, born in 1855, aimed for a successful career like that of his father, who had sailed the full-rigged *Jawa,* up to his death at 40 on the solitary island Saint Helena.

Accordingly, the son was beside himself of joy when he at 15 was employed as apprentice on uncle Axel Söderström's *Thor* between 1870 and 1872, during which period the teenager encountered the exciting

---

[4] Gruesome with a vengeance—as when J. Paul Getty received his son's ear by post in a cardboard box.

City of London with its lures of 1870, and Kingston, Jamaica, in 1871. As it happened, in both ports he met with an old schoolmate, Rickard Garberg, who was deckhand on the full-rigged ship *Sophie*, which for a long time accompanied the *Thor* in these harbours, both being freighters. There were extended periods of loading and discharging, so Pehr Johan joined Rickard ashore on the first Sunday, the day off. Garberg tells the story in his memoir *I rigg och skans* [Garberg, 1936].[5]

They asked captain Söderström for permission and a shilling, which were both granted, while the captain—'a kind man and John's uncle'—with a laugh warned them of the Sunday Girls. As nobody else was interested in seeing the sights of the city, the two went off alone. They followed a signpost saying 'London Bridge'—which was very far from the dock. They went on endlessly, as it seemed, through deserted areas, until they were suddenly besieged by a swarm of unabashed girls who made passes at them, clung on to them, kissed them, whispering seducing words, which they could understand after having studied English for a year or two at elementary school: 'My darling, come along!', 'You will come this evening, won't you?' and other things to the same effect. They fought to get away and shake them off; there was a brawl, and the police was sent for with a whistle—who laughed at the puerile sailors. They never got to London Bridge but were relieved to get unscathed back to the ships.

In Kingston next year, where the *Thor* had come with ice cargo from New York, the ship decks were crowded by black dancing-girls in the heat, whom Pehr Johan joined in the dance. Thereupon the dancers enjoyed seeing the young seamen dive from the yards and escape the sharks in the breakers. But the deepest impression on the two young men was made by seeing a cutter with naked and emaciated convicts in the harbour. The captives all had letters and numbers on their backs and were swayed by officers with knouts dealing out whizzing blows to the howling slave labourers.

After these years on the *Thor* with good service record, Pehr Johan was thrilled when he was admitted as ordinary seaman on the full-rigged

---

[5] Pehr Johan Zellinger, mostly called Johan, is named John Zellund in that book, entitled 'In rig and forecastle,' which has the form of a *roman à clef*. Rickard Wilhelm Garberg was born in 1854, and one year Pehr Johan's senior.

Superior, navigated by his other maternal uncle, Anders Clase. There he got advanced training in entering the rig, hoisting sails, and serving as lookout in the crow's nest.

Taking in sails at the port and goods terminal up in Ljusne, he unfortunately fell to the deck from the main mast and broke his spine. His life was beyond saving and he gave up the ghost at the age of eighteen on 10 November 1873.

His mate Rickard remembered: 'He was a lively fellow, like me, and a good sport. Unfortunately, a few years later, he made a somersault from the main yard, whereupon he crushed his back against the frame of the hatchway and suffered an instant death' [Garberg, 1936, 29].

An urgent letter with the sad news was written to his mother Charlotta at the home address and dispatched by mail. The corpse was put into a wooden chest, which was carefully nailed-up and sent by boat to the home port, just 37 nautical miles away—with the same addressee.

The boat was probably a steamer, and the envelope lagged behind. The bane of the slowness of postal service was never worse than in this case. The coffin arrived to mother Charlotta without previous warning. The body preceded the news of her son's unexpected death.

Charlotta never got over the shock. Afterwards, there was also a rift in the relation to her sister Johanna Clase and her family. They did not get along, and finally did not talk.

There was other delayed mournful news. This was a grief-stricken year for the families. Her brother Axel did not learn about what happened to his own family until he returned with the *Thor* from the Caribbean and learned that he had become a widower and his children motherless.

Amalia Mathilda had passed away on Tuesday, 18 February 1873.

She was one of the many victims claimed by tuberculosis, a common illness in nineteenth-century Europe. Like cholera, it is a bacterial disease, but a contagion spreading from one person to another, and usually affecting the lungs. A weakening affliction, and, in the exacerbated pulmonary condition, slowly suffocating the feverish blood-spitting sufferers.

Before Amalia Mathilda Söderström was buried the following Wednesday, on 26 February, this 24-year-old woman lay in state at

Fig. 3. *Amalia Mathilda lying in state in a ship coffin in 1873.* The Maria L. Hallengren Collection.

church and during the funeral service, visible high on top in a boat-shaped coffin with her face turned toward the congregation; her eyes were shut but her mouth was open (Figure 3).

The sarcophagus and all arrangements had been made by mariners, shipbuilders, widows and friends, according to an ancient and almost forgotten Northern custom. It answers to the archetypical form found in prehistorical stone ships, a symbol with universal signification throughout the ages. The voyage of the dead continues. Ship models were the most common votive gifts in seaport churches, and so it was at Holy Trinity Church, as elsewhere. The Ship was a symbol in ancient and medieval Scandinavia, and to the seamen it remained as sacred as the Holy Virgin.

The two daughters Maria, 5, and Emmy, 4, were present in the church, and watching their mother and all people slowly passing the body in tears. Some they recognized, others not. They particularly noticed a sobbing lady in mourning veil, whom they could not identify and who never made herself known. Their grandmother was dead, and so was the great-grandmother. For life the sisters wondered who the inconsolable mourner was, dressed in black and with a crape. A key to their origin and life story?

There is every reason to believe that this was the close friend and cohabitant of their grandmother, Johanna Lindberg, who took care of the young Mathilda and Carl Johan at Carolina's deathbed. This can be concluded by elimination and makes sense. Why did she not approach them, then? Maybe, she did not know them or the others present at Mathilda's funeral service, or perhaps her social standing prevented her from interfering. We will never know.

There was some hush-hush about these matters, and a smoke-screen had settled upon the past, particularly after the orphaned Mathilda was adopted and sent to school. There was a wall between social strata in the most insignificant societies. When Maria and her sister Emmy were small, they had once been playing outdoors when a cleaning lady or laundry woman suddenly addressed them and told them that she was their grandmother's sister. They had not believed her but rushed in to their 'grandmother' substitute Mrs Winroth, and told this, whereupon she silenced them and said that this was nothing to talk about.

Nevertheless, the incident had obviously made a strong impression as Maria remembered this—and was to tell her children and grand-children about it. The young ones were in a state of uncertainty about these things and were kept in the dark. They had no acquaintance with their real grandmother's three sisters, two of whom were actually dressmakers, and one who was married to a tailor and junior schoolteacher and gave birth to fifteen children—in fact named Thilda Löfström like their mother! They would never know. There were watertight bulkheads as on the sea. However, conceivably some of these people attended Mathilda's funeral and were in the row to the coffin.

After Mathilda's tragic death, the one who now took over the maternity role was—once again—the motherly Mrs Winroth, the founder of the orphanage who adopted the charity girl Mathilda, their mother. There was a recurrence in their fates. This 'Grandma' was to remain their safe refuge and centre of tenderness, and later on the lasting object of Emmy's and Maria's love, longing, worries, and care—until she died in 1893 at the age of 82.

7. The Importance of Learning

The tragedy did not pass without notice. Another one who stood up for the girls was a headmaster's wife, Mrs Clara Klintberg, who was thereafter always called 'Aunt Clara' by the two sisters—even though she was not a close relative. Why was she so seriously concerned about their well-being and committed to their cause?

Families took care of one another. In a letter to her son Carl, who then studied Law at Upsala University, Clara Klintberg wrote on 21 February 1873:

> 'Now, little Carl, Thilda Söderström's suffering has ended. Tuesday at noon she departed this life. A few days before, she was indeed weaker than usually, but no one could believe that the end was so near. Poor Charlotte W. and the small children; [Captain] Söderström is in the West Indies, and for a long time will not know about their loss. She will be buried next Wednesday.'

With Clara, who cared so much for the departed and her daughters, there was both an affinity, an important school connection, and a vibrant awareness of death.

Clara Klintberg (1822–1885) was the daughter of the municipal medical doctor Carl Nordbladh. Since 1851 she was married to Anders Gustaf Klintberg (1816–1884), a legendary figure in nineteenth-century Gefle. He was the principal of the high school from 1866 to 1884, and was a man who instilled respect in both teachers and students. The discipline was total.

There was a merchant in Gefle named Nils Gustaf Osenius, a relation of the aforementioned skipper Anders Osenius. Anders and his wife Brita Elisabeth Hammar had a daughter who was married to the factory inspector A. U. Leufvenius in Ockelbo, and they in turn had a daughter, Charlotta Elisabeth, married to the sea captain Olof Winroth (1803–1855) in Gefle.

Nils Gustaf was married to Margareta Rosenbaum (b. 1799), and they had a daughter Emma who was born in 1828. Margareta became a widow when the daughter was small and remarried in 1833 with Dr Carl Nordbladh. He had become a widower the same year and was left alone with three daughters, Carolina (b. 1814), Clara, and Elise (b. 1829). Their two older brothers had emigrated. Carl and Margareta and the four girls lived in farm No. 92 in Gefle City's Second District. It was located a hundred yards northwest of the church. Carl passed away on 27 July 1855.

In this century of afflictions, unreliable death criteria and emergency interments, municipal medical doctor Carl Nordbladh had taken due precautions. For very good reasons he feared to be buried alive in suspended animation—which occurred from time to time. The doctor knew. In his will of 1846, he had stated:

> 1. After I have passed away from the noise of this troubled world, my corpse should remain on the deathbed for a day if it is winter, or for half a day if in summer time; thereafter taken up and wrapped in the sheet on which I died, and should then remain so lying until cadaverous odour is noticed; then it should be placed in a coffin filled with chopped spruce twigs, and, if it is in the summer, regularly sprinkled with chlorine water, without any winding sheet. 2. The burial should then take place as soon as the circumstances permit, but not until the odour is perceived.

The importance of Aunt Clara and her husband, the headmaster, is crucial. Together with Charlotta Winroth they see to that Maria and Emmy are taken care of and are well brought up. To them all, schooling is the next priority, and they will arrange for this; first in nursery school and junior school, then in an eight-year private girls' school. Several enterprising women in town are working on these things on behalf of the education and liberation of women—a new idea! Consequently, there are future opportunities and their prospects good.

However, due to the fact that paternity was so often undetermined; seaport fathers so often dead or absent; and since 'town intelligence' was the regular coffee table news-service, all this care of the elders as regards tutoring and upbringing inspired the tittle-tattle tell-tale that Emmy's and Maria's grandfather in reality was one of the Gefle headmasters. The prominent principals Karl Johan Moberg and Gustaf Klintberg were the hopeful guesses, both born in the 1810s. The grapevine knew that Clara's husband had given a maid in his parental home, a certain

Ms Zetterström, a child. The offspring was subsequently given away, while the mother became distressed and abided elsewhere, until she re-entered upon duty with the renowned MP Edvard Casparsson at the Brunna manorial estate in Upland. That was the story.

By and large, this host of rumours did not arise from spite. Uneducated citizens wished to dispel uncertainty and ambiguity, and set all things right. Children also wanted to know. The one conclusion that can be drawn from the overwhelming torrent of gossip in this era, is that paternity was considered sacred and the present state of affairs unholy. The groundwork of society was in the balance, and the ignorance of the poor considerable. In this case, the fact is that all the girl's mentors and benefactors were committed women, dedicated to the cause of girls.

For this reason, the children could thank their lucky stars. With such benefactors they had every chance of succeeding. Furthermore, the estate of the deceased mother showed that there was no lack of means. The inventory was made by the body of borough administrators on 14 June 1873, when the widower had returned home from the sea.

On the debit side it can be observed that the funeral was indeed lavish: 300 riksdaler, which was a lot of money, considering that an industrial worker earned 1.50 rd. a day.

The comprehensive deed, which results in a balance of 1457 rd., lists, among many other things: a pocket watch, rings, brooches, and earrings, all in gold; silver cutlery; one and a half dozen dishes and five dozen plates of porcelain; one dozen wine glasses, one dozen punch glasses; coffee set with two dozen cups and saucers; two black silk dresses (40 rd.); three wool dresses; winter coat, summer coat, and a Bedouin coat; three wool shawls of fine quality (30 rd.); a lace shawl (20 rd.); umbrella; parasol; two sofas and ten chairs with covers (100 rd.); two tables veneered with mahogany; the piano (20 rd.)[6]—under which they huddled up during the terrible night of 11 July 1869!—a

---

[6] Twenty riksdaler in coins answered to 0.5 kg silver. This year, in 1873, gold standard was introduced in Sweden and the monetary unit was changed. Riksdaler (rd.) was converted to Crowns (krona) set at the same value, and in this new coinage 1 rd. = 1 kr. 20 kr in 1873 corresponded to about 1,000 kr in 2021. It should furthermore be observed that at the winding up of estates the estimates of value were invariably on the low side.

wall clock (20 rd.); and, along with the beds and beddings—including the mentioned wedding sheets!—a touching cot.

Apparently, the wedding presents of 1866 included not only the fine linen, but also the musical instrument, which once served as the roof over the head of toddlers.

8. The Great Liberation

'An educational establishment for all time'

The momentous Enlightenment concept of *human rights* became a revolutionary force in the nineteenth century when it was realized that the human species included all people on earth, all social strata, and women as well! Assuming the challenge and responsibility, enthusiastic elders and idealistic woman teachers founded girls' boarding schools, which grew to become public schools for rich and poor—with freedom, equality, and sisterhood as guiding principles. The main driving force was a belief in growth and liberation by education and edification. In Gefle, this happened when Emmy and Maria reached preschool age.

Foremost and most radical among the headmasters, Karolina Själander considered the education of girls more important than the education of boys, because the mademoiselles would be the mothers and educators of the next generation! She believed in progress, knowledge, and heartfelt understanding, not in lessons learnt by heart.

The motto of the curriculum in her pivotal girls' school in Gefle was *non multa sed multum*, 'much about little and not a little about much.' This entailed long school days and heavy homework, as the syllabi contained Euclid's *Elements of Geometry*, textbooks in algebra and arithmetic as well as botany, zoology, chemistry, physics, and modern languages—along with the timetabled writing of compositions; music, art, and gymnastics lessons.

The good tone was tuned in and set by the disciplined forms of the school. When the '*Tante*' entered the classroom with her briefcase, she was greeted with a song. The aim, motto, and guiding principle was: 'An educational establishment for all time' [Själander, 1941, 14].

To Själander, lessons without a teacher was like a reader without a book, and to become a member of her staff the requirement would be

Fig. 4. *Karolina Själander (head mistress) and Klara Johansson (home room teacher).* Courtesy of Gävle Länsarkiv. A unique photograph: Headmaster Karolina Själander (standing) with form mistress and scientist Klara Johansson (sitting), who forever left their marks on the Söderström daughters. The Wilhelm Lindeberg Portrait Gallery. Photographer: Carl Berggren (1847–1897). Länsmuseet Gävleborg, public domain. XLM.U06867.

five years of practice after teachers' training course for tenure. As homeroom teachers, the most dedicated were called to provide continuity and harmony through all forms and hold all things together, attending their pupils for almost a decade in their development and formation of character.

For this reason, Maria Söderström was to see the optimist '*Tante*' Klara Johansson a lot, the devoted naturalist dressed in black and with her dark hair parted in the middle as the headmistress' (Figure 4). Ms Johansson contradicted all complaints saying that all weather is God's weather and every flower on the ground a miracle. Her sprightly motto was: 'rather suffer injustice than being unjust.' She was remembered as just and humorous.

There were numerous forerunners in the town. Bonne Amie Flygare's junior school, the boarding school of Bonne Amie Sundström, and that of Councillor Strömbäck's daughters; the private school of Cecilia Fryxell, who also trained teachers, as did Sunday school instructor Sophie Garberg. There were the remarkable girls' school founders and headmistresses Elsa Borg (b. 1826), and Clara Lind (b. 1841).

Karolina Själander in succession took over two other schools when Borg moved her activity to Stockholm in 1874, and Lind relocated to Upsala in 1879. In the early years she had moved her protégées to her old parent's coastal home for more space. After the fire of 1869, it had been hard to find suitable premises. The new classroom was in the ambience of boathouses, hamlets, ropewalks, artisans, fishermen, seamen, the fresh sea air of the north—her origin. Her brother Per Martin was the captain of the brig *Maria*, which carried emigrants to the USA. In 1841 their parents had brought the new-born Karolina from the fishing season up on the Norrland coast. Since then, their father had become a wealthy businessman and joint owner of sailing ships, and the stone house in Gefle was roomy; and is still standing.

The school expanded, and was renamed and moved. It would slough its skin from elementary school to *Higher Girls' School* and become a large establishment, which ran an eight-year elementary school and a three-year preparatory school. Using all her heritage, selling her home, and on her own risk, she finally had a new schoolhouse built in the town to meet the needs of the community. It was inaugurated in the autumn of 1878 and became known as Själander's school.

II. Maids, Orphans and Women's Rights

It was a magnificent renaissance edifice in the town hall Esplanade, consisting of two floors, with shop premises for rent on the ground floor; seventeen classrooms, reference library, prayer hall and chemistry room on the second floor; and attic flats above, where the principal Ms Själander lived under the ridge as a spider in a web (Figure 5).

Boys in the street always cast a covetous eye in the direction of the closely guarded school-building where Själander ruled, calling it 'the roost.'

She never married and remained the *tante* who looked after all the girls below, along with taking tender care of nephews and nieces under her roof. She gained celebrity status, notoriety with some, as an ardent suffragette, YWCA worker, nonconformist, and vegetarian. Pupils also remembered the dairy and toffee vended on the street level.

Fig. 5. *The Higher Girls' School, the magnificent creation of Karolina Själander.* Courtesy of the Länsmuseet Gävleborg. The Gustaf Swedlund Collection. Creative Commons Share Alike CC BY-NC-SA. XLM. U09787.

Fig. 6. *Emy (upper left) and Mia (second right) with fellow-students and kinsfolk in 1884 before departure. The boy is their cousin Knut Zellinger.* Photo Elise Ahlgren Gefle. Private possession. The Hallengren Archives.

On the darker side, the demanding standards and the almost unattainable level of ambition; desk-replacing in order of precedence each week; duty calls and reproaching hellfire sermons in the prayer hall, on top of insulting reprimands and disgraces, implanted a sense of sin and shame and a consciousness of guilt along with profound knowledge and sturdy integrity. *Non scholæ sed vitæ discimus* 'we do not learn for school but for life' was a hackneyed maxim put into practice in all late nineteenth-century girls' classes, for the reason that life was hard and inequitable, and schooling a survival course for the fair and weaker sex. Aiming for the stars, it sometimes overshot the mark or hit the bushes.

A grateful posterity, including women's lib and the number of successful students, would recognize the value of Ms Själander's disinterested pioneer work, and her pivotal role in the educational system. In 1901 the government awarded her the royal gold medal *Illis quorum meruere labores* ('for those whose labours have deserved it'). Since 2020 she is one of the eighteen national celebrities honoured in the local Gävle Walk of Fame.

Thanks to all these initiatives and their benefactors, and their absent father paying the reduced term fees, Maria and Emmy Söderström were finally admitted to the eight-year elementary school in 1878 [*Redogörelse*, 1884, 31] (Figure 6).

The sisters did not see their father for years, since he had become the master mariner of the new gem and pride of O. A. Brodin's wharf, the 187 feet barque *Atlantic*, launched in 1876 (Figure 9).

In the 1880s, he was at home to pick them up from school only once—at graduation day in 1885, when they were free and permitted to sail, and then he brought them on a trip round the world (Figures 7-8, Ch. III).

III. The Travel Diary

'Notes during the Sea Voyage

7.11.1885–25.9.1887'

By Maria Söderström

The first diary from a circumnavigation written

by a North European woman

—published for the first time—

With asides in italics and footnotes.

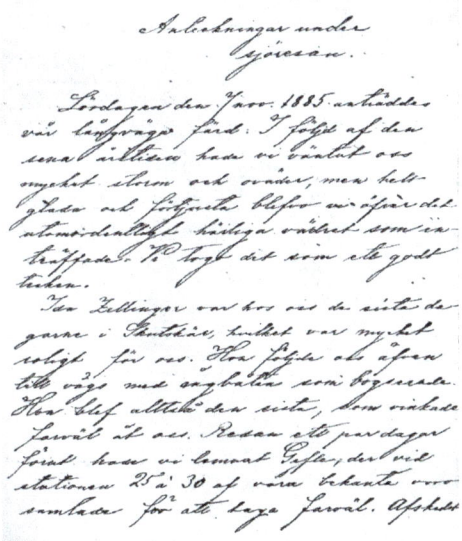

Fig. 7. *Maria Söderström's travel diary on a table.* Private possession. Author's photo.

Fig. 8. *The journal begins. The first page of the travel diary.* Private possession. Author's photo.

Fig. 9. *'THE ATLANTIC. Commanded by Captain J. A. Söderström. Oil by H. Petersen & P.G. Holm, 1877.'* The barque with all square sails and for-and-aft rigged mizzen sails set. The efficiency of the fast-sailing merchantman in various weather conditions and winds could be improved by extra gaff sails, trysails or staysails including jibs. She is carrying the Swedish–Norwegian Union colours along with flying the HNJB signal flags, the ship's identification (Figures 21 a-d, Ch. IV:12). Private possession. The Hallengren Archives. Author's photograph.

1. From the Gulf of Bothnia to Øresund and Zealand

Departure at Port of Gävle (Gefle)—Farewell to the loved ones

'ON SATURDAY, NOVEMBER 7, IN 1885, WE SET OFF UPON OUR LONG JOURNEY! Due to our late departure at an advanced stage of the season, we had expected a lot of gales and storms, but we were immensely relieved and delighted by the extraordinarily glorious weather conditions as we set off. We took this as a good omen.

[Cousin] Ida Zellinger[1] was with us the last days in Skutskär,[2] which was very fun for us. She also followed us along the way with the steamboat that was towing the ship. She thus became the last to wave goodbye to us. A couple of days before we had left Gefle, where at the station 25 to 30 of our acquaintances were gathered to say goodbye.

The parting from Grandma[3] was especially sad; alas! The dear beloved grandmother—how heartbreaking to think that we may never again on earth see her again! I cannot hold back my tears at the thought of her last farewell: "If we no more meet on earth, know for sure: once in heaven!" Yes, we can believe and hope so. With heartfelt prayers to God, I have begun this journey, alas, the faithful God probably wants to hear my prayers and let everything go well.

Adage for the day and for the journey: Jesus Christ *is* the same yesterday and today and forever. Hebrews 13: 8.

MONDAY, 9 NOVEMBER, 1885

We are now off the coast of Gotland, and so close that we can distinguish buildings and church towers. It has gone pretty fast. This is due to the favourable winds during the daytime yesterday, which also

---

[1] Their elder cousin Ida Katarina Zellinger, born in 1861.
[2] Skutskär (meaning 'Ship Skerry') at the Northern Upland coast is an old anchoring place for the iron and wood export industries in the region, and a bi-municipality at the mouth of the Boda and Dalälven rivers by the Gulf of Bothnia. The *Atlantic* regularly took in commercial cargo there, so to boot for this foreign journey!
[3] Their mother Mathilda's foster mother Elisabeth Charlotta Winroth, née Leufvenius, (1810–1893), married to sea captain Olof Winroth. Their biological grandmother Carolina Petronella Löfström died in the cholera epidemic of 1853, leaving her five-year-old daughter an orphan. Father unknown.

caused the seasickness to appear with us as an expected but unwelcome guest. *Huh!* how painfully sick we were, but fortunately at least transient. Already last night I began to feel hungry, and with increasing appetite we ate herring and potatoes for breakfast.

Today it is the most glorious weather of all time, almost still, so that we do not get very far today.

The thoughts return again and again to the loved ones at home. Alas, of course they are thinking of us even now. Especially Grandma, every moment she remembers us and prays to God for us. Poor grandmother what she should feel worried about us and long for us, but no less we for her!

Tuesday, 10 November, 1885

Delightfully nice day! I'm starting to think that it can impossibly be tedious or unpleasant on the sea, at least not on the *Atlantic*, in the company of our dear, dear father!

The day has been spent with work, reading, music, walking, etc. The weather is the most wonderful you can imagine. "Unparalleled weather in the month of November", I hear Dad exclaim! Steersman Elfström becomes seasick in weather like this, he states. He probably just dislikes the fact that it does not blow enough for it to go away well. Nevertheless, we are now already at Öland's southern cape.

Best of all, today I have felt so deeply connected with my God and Saviour. I have read an extraordinarily beautiful consideration from the little book *Abide in Christ*, which I bought on the recommendation of little Hedde.[4]

It is especially about the abode of the crucified Christ. It showed how necessary it is for a Christian not only to regard the cross as the basis of our salvation but also to take on the whole crucifixion of Christ. Words worth considering! O Jesus, teach me to remain and walk in the imitation of you.

---

[4] She is referring to the pamphlet *Förblifven i Kristus!: Tanken öfver det saliga lifvet i gemenskap med Guds son* (Stockholm: C. A. V. Lundholm, 1884), a Swedish translation of *Abiding in Christ* (1882) by the South African writer Andrew Murray (1828–1917).—'Hedde' is a schoolmate from the Själander Girl's School in Gefle, Hedvig Katarina von Post, born 1868, a close friend of the sisters who is often referred to.

Wednesday, 11 November

Sunlight by day, starlight and moonlight by night: such has so far sea life seemed to us!

It is with real delight that we enjoy the glorious weather, and this on the Baltic Sea in winter!

Today it is one week ago that we said goodbye to Grandma. Although each day has passed very fast, I still think it is much longer than a single week since then.

> Our loved ones we lay
> Dear Lord for your foot
> Trusting your promises,
> That our prayer you receive,
> And where would we else turn in prayer
> Than to the Father who fosters us
> For our right home.
>
> And though days quickly pass
> And though friends part here
> Are you, however, the association band
> Between those we love.
> May we therefore in all shifts
> Of joy and trouble
> Faithfully hold on to Him
> He who is eternal love.

Elsinore [Helsingør], 17 November

Last night we arrived here. The *Atlantic* is now at anchor and waiting for good wind. Yesterday we were ashore and visited *den smukke by Helsingør*, "the beautiful town Elsinore". Not that it is really very "beautiful"! It would probably have been more enjoyable to get up to Copenhagen, but the time was too short. We visited a family named Schierbeck.[5] They had a cute and lovely little daughter. We have now also sent letters to our loved ones back home.

---

[5] Schierbeck is a Danish family that is famous for its musicians and artists.

It's been quite a few days since I last made notes about the trip. For two or three days we had stormy weather, and we poor girls were quite seasick. May the journey through the North Sea now go quickly, so that we do not suffer too much. However, I hope that we become more and more hardened, and in the end really good seafarers [*sjögastar*]!'[6]

2. The North Sea

'THE ATLANTIC VOYAGE, 10 DECEMBER, 1885

What a punctual set of diary notes I keep! Nearly a month has passed since I last opened this book. In a few words, I would like to summarize how our journey during this long time turned out. It has for the most part been very pleasant and varied, and everything has gone happily and well. In the North Sea we had excellent good weather, beyond all expectations!'

As all other sailors, the captain knew very well the perils of these North Sea waters. Here, his elder brother, the sometime captain Carl Gustaf Söderström of the brig JUNO, *perished on 9 October, 1856, aged twenty-seven.*

'In the ENGLISH CHANNEL, on the other hand, it was several days nasty and stormy—there we went through our last ordeal of seasickness, as I hope!

---

[6] Mia's use of the Swedish word *gastar* (about 'men' on board), when she refers to herself, her sister and the crew as *sjögastar*, was partly inspired by the derelict song in Robert Louis Stevenson's *Treasure Island* of 1883 'Fifteen Men on a Dead Man's Chest', soon sung by sailors in their daily chants and translated into Swedish as 'Femton *gastar* på död mans kista'. This became so popular that it occurred in Astrid Lindgren's *Pippi Longstocking* and in *Pirates of the Caribbean: Dead Man's Chest* in a later era. Thus, words originally stemming from a book by Charles Kingsley on Dead Chest Island in the British Virgin Islands, termed Dead Man's Chest, echoed for centuries and entered Swedish usage and was known by seamen on the *Atlantic*.

Praise be to God, who so lovingly helped us avoid all dangers. We are now at about the height of GIBRALTAR. The sun in Spain shines so glorious and clear, it's like the sweetest mildest spring weather with 20 degrees Celsius in the shade. I hope that our winter is now largely over. Oh, how nice! It is now only two weeks until Christmas. What time goes fast!

It will probably be a lot of fun to celebrate Christmas here on board with our beloved dad.

It is now also ten years since the last time we had dad with us at Christmas!

11 DECEMBER, 1885

Memorable words from the sacred text have today made my mind solemn and thoughtful. Rather, through a little book I read, the Holy Spirit of God has graciously shown me that for a Christian who wants to remain completely in Jesus, the Saviour to whose guidance she completely wants to submit to, it is especially necessary to consider that the whole life of the Christian should be a testimony of the eternal faithful love of Jesus, who sacrificed himself in death, and whole-heartedly devoted himself to serve, help, and to forsake all things for the brethren. O Jesus, my dear Saviour, you see my heart, my arrogant, vain and proud mind, Oh, how can I be a testimony of you before my fellow men, even for my loved ones? My tied tongue has nothing to say about you and your love, although the desire of my heart daily is to serve, honour and praise you. I lack courage, strength and determination, yes, Lord, I lack everything without you, but if I am weak you are strong, then be my strength, faithful Saviour, and teach me to embrace you completely as my help, my comfort and my bliss.

19 DECEMBER

Bless the *Lord*, O my soul; And all that is within me, bless His holy name!
Bless the *Lord*, O my soul, And forget not all His benefits:
Who forgives all your iniquities, Who heals all your diseases,
Who redeems your life from destruction,

Who crowns you with loving kindness and tender mercies.
(Psalms 103: 1, 2, 3, 4)

See in this the fundamental tone of my heart's mood this day! Yes, the Lord God is the only one who can fill the heart with peace and joy. I feel so deeply happy and fortunate now in the knowledge that I belong to Him.

Lord Jesus, I want to trust in you forever. I want to completely sacrifice myself for you, I want to belong to you with body, soul and spirit, then give me grace to be able to serve you in holiness and righteousness always, and let me sense your presence, and give me your peace, O Jesus!

For a few days now, we have been sailing in the lovely NORTH-EASTERLY TRADE WIND. The day before yesterday, we passed MADEIRA. We can now, I hope, expect excellent weather in the coming weeks. We now spend whole days out on deck, busy with our work, and furthermore with our English lessons, to which two hours are usually spent a day. Let's see how persevering we will be in our studies! I spend long hours of the day aloof by the railing, immersed in the glorious sea; how lovely it is to listen to the sound and surge of the waves, yes, nothing I think, can in sublime grandeur be likened to the vast immeasurable sea, above which the blue sky rises, especially today, so bright and clear. Only a few bright, light clouds hover on the horizon.

It is now near sunset and between these clouds, which here and there change to purple, the sun descends so lovely and radiant. A sunset on the sea must unconditionally give the mind a peaceful atmosphere.

> What does it matter that the twilight widens
> And stretches out its veil,
> She is the friend of thoughts.
> What does it matter that the sun sinks,
> She's returning tomorrow again.[7]

---

[7] She is here quoting a verse written by the Swedish poet Karl August Nicander (1799–1839).

This also applies to God's grace! It never goes away, even though it often appears to do so. This thought should comfort and enliven the heart. If only I could hold on to that! God help me with this!'

3. Under the Tropic of Cancer

'20 December, 1885 [Sunday]

I want to note this day as one of the most peaceful Sabbath days I have ever experienced. Today I have clearly felt the presence of the Lord. Yes, "He is close in his Word", as it says in the little beautiful song I remember my dear friend Hanna P.[8] often used to sing. I have spent much of the day praying to God and reading His word. A beautiful Advent sermon by Waldenström[9] clearly pointed out where we find our God and all His love revealed: "Well, look down to the child in the manger, there you see God; look up to the Lamb on the cross, there you see God's thoughts and will."

Now that Christmas is getting closer and closer, the thought goes back more and more often to the loved ones back home. We now know so clearly what they are up to: today in particular is Father Christmas' annual party, which we have attended every year; I think the young ones of Zellingers'[10] are there this year too. Besides, they are obviously now busy with getting the homes nice, assembling decorations, and all sorts of preparations for Christmas, such as Christmas gift

---

[8] Hanna Pettersson.
[9] Paul Peter Waldenström (1838–1917) was a prominent leader of the nonconformist Free Church movement in late nineteenth-century Sweden.
[10] The reason why the Zellinger and Clase family names got so closely involved in the Söderström life story was that these were seafaring families interconnected by marriage. Captain Axel Söderström's sister Johanna Katarina Söderström (1825–1915) married captain Anders Leonard Clase (1814–1898) in 1845, and his other sister, Catharina Charlotta Söderström, in 1852 married captain Pehr Zellinger (who was born in 1823 and killed at the tropical island of St Helena in 1864—their son, the sailor Pehr Johan Zellinger fell from a mainmast and died at 18). Thus, it happened that the names of Clase and Zellinger appeared among the crew of the *Atlantic* during the long life of this enduring sailing ship, and even as mate or master. The ill-fated Pehr Johan's younger brother Knut Zellinger served as the master mariner of 1901–1902 before his early death at Lake Michigan in 1903, aged 41.

arrangements, etc. Our beloved grandmother, alas, she will probably miss us a lot during Christmas! I wonder how it is with her now, oh, may she be healthy and well, and live many more years, so that we may see her again when we come home, therefore I pray daily to God; yet may in all *His* will be done!

23 December, 1885 [Wednesday]

Today we have passed the Cape Verde Islands. One of them, St. Antonia [Santo Antão], we saw very close. It was admirably beautiful, with sky-high granite mountains that go sheer down into the sea.

I do not know why, but in recent days I have not been in a mood suitable for Christmas, which is now so imminent. I have been distracted in many ways; the thoughts have not been able to gather in prayer to God: in short, I have been restless and anxious. However, I hope that during the beautiful Christmas holidays, the Lord God wants to be especially close to me with his grace, peace and spirit.

There are three holidays in addition to Christmas Eve tomorrow. It will be nice to rest the heart in God's word during these days.

Tomorrow it will probably be fun. Then we may open the Christmas parcels that have been sent for us from several of our friends back home. I have been quite curious to see what they might contain, letters and probably small gifts.

Preparations for Christmas are being made here, I have noticed from our nimble and able steward Östling. He is busy baking several rare pastries, with which he wants to surprise us all, I can believe.'

Carl Östling, cook and steward, born in Gefle in 1848, a 37-year-old married man, with family left alone ashore all the year round as other sailing husbands. Emy and Mia often assisted him as hands in his daily work as washers, cleaners, cold dish manageresses and servants, since the ship lacked the usual 'cook boy' and messroom boy, and they felt at home in the warmth of the galley, especially when they travelled far in the Northern and Southern Seas. They were with him in his work on deck but not in the rig—the use of trousers was still restricted to men. A skilled and experienced chef, the ship's cook worked miracles, producing as if from nowhere tasty meals. Sea air gives an appetite, but perishables were rare, and the food

supply was poor and limited before the era of freezers, refrigerators, and the growth of an international tin industry. The fact that Östling had picked up some British cookery for seamen was an asset, along with his economical fattening of domestic animals on board with leftovers, peelings, refuse, and dishwater—detergents were never used on board.

'It looks like we'll have nice and wonderful weather this weekend, for which I'm sincerely grateful to God, for I have asked Him for it. A comforting word from God came to me right now, which I want to state here and thus end today's notes: "In returning and rest you shall be saved; In quietness and confidence shall be your strength" (Isaiah 30:15).

Lord Jesus, teach me to be still before you and trust in you with all my heart, and thus be helped and strengthened, as you promised here in your word. Amen.

CHRISTMAS EVE, 1885

I now go on to describe, if possible, the day about which Emy and I have agreed upon is the nicest and most peaceful Christmas Eve we have ever experienced!

A true Christmas peace has reigned in my heart throughout this memorable day.

I have very clearly felt the Lord's presence, and am dumbfounded at how big His love to us poor sinners must be, as He has sent His only beloved Son down into the world to be born as a little child in poverty and meanness—"that by His poverty we shall become rich". Today is born to us the Saviour, who is Christ, the Lord in the city of David.'

4. Celebrating Christmas on the *Atlantic*

'Already at dawn and in the broad morning light everything appeared to have a festive touch. Maybe it was the extraordinarily beautiful weather that produced this to some extent: clear, bright skies, mild summer-like air, glittering blue waves. Everyone seemed happy and content, yes it was clear: Christmas is here!

After we had breakfast, Emy and I could no longer refrain from opening the parcels we had received from home.

To our regret, I have to write this down, we were missing something from Grandma. Alas, the dear beloved grandmother, how I miss her, and how much I long to be able to embrace her and tell her so! However, she had so much to think about and worry about at the time we left, that she did not have time to think of sending us a remembrance at Christmas time, but she probably wish us well anyway!

Instead, there were long and friendly letters with several small gifts from our friends Lully, Hedvig v. P. and Hanna P.[11] How fun it was to see their assurances that especially during the Christmas days, when we are so far apart from each other, that they are all thinking about us, and each pray to the faithful God that our journey may go well and hope for a happy reunion.

Several pleasant surprises awaited us in the evening. We walked with father at quarterdeck and had no clue of what was to happen, when the meal bell chimed. We went down to the dining saloon, where the Christmas table was set "in all its glory". At the top of the table, a stately branched candlestick, adorned with gold and silver paper and in blazing colours of the season, had a rather close resemblance to the Christmas tree, or rather reminded of it, at least, dazzling as it was with candles along with all sorts of "yum-yum", such as apples, little Christmas goats, goblins, etc., in cake pastry.

During the meal where we, sure enough, were treated to fish and porridge according to the custom, we heard a strange murmur, and a trudging back and forth outside the cabin doors, and from time to time perceived the helmsman's commanding voice. We went out and were not little surprised to see that the deck up amidships had been transformed into a large hall, richly adorned with flags and coloured lanterns. The crew just put the finishing touches to the work when we came out.

Dad and the first mate wished them all a "Merry Christmas", and one of the crew responded with thanks in the name of all of them, after which they went midships, where they were treated to punch, cigars etc.

Throughout the evening and well into the night, they sang, played music, and danced; perhaps an unsuitable way to celebrate Christmas

---

[11] Lully Levin, Hedvig von Post, Hanna Pettersson.

Eve, the commemoration of the birth of Jesus, but, sorry to say, for many this was not the main subject of joy this day.

Holy dear Saviour, make yourself truly known to more and more of us poor people, who without you are nothing, but with you get everything, life, peace and bliss. "Draw us to you by your sole mercy," as you have drawn me to you and made you so absolutely indispensable to my heart. So, let nothing be able to pull me away from you, but instead let everything that happens to me, joy or sorrow, bring me ever closer to you!

> Closer to Jesus Christ, to you,
> Closer to you whatever it costs me,
> Closer tomorrow than yesterday,
> Closer, O Lord, from year to year![12]

CHRISTMAS DAY [25 DECEMBER, 1885]

For much of the morning, Dad and both of us were busy reciting the sermon of the Christmas Gospel on board with the crew (Figure 10). This was nice and refreshing.

Alone again in the early afternoon, I finished an excellent book, which I had been reading for some time. It was called: "On the way of the cross to the crown of life", and gave full reason for its name, because it clearly showed how several crusaders, men and women, happily reached the goal and won the crown after a battle.[13] Actually, it was one of the best books I have ever read, and it made me very happy to study it.

Towards the evening we had a lot of enjoyment by seeing how the full crew amused themselves with all kinds of games, pranks, and jokes.

---

[12] Hymn by Lina Sandell (1832–1903), female Swedish poet and author of gospel hymns. The text set to music by composer Oscar Ahnfeldt (1813–1882). It was published in *Samlade Sånger af Lina Sandell*, Vol. 1, Stockholm: Ivar Haeggström, 1882. The sisters sang this song on Christmas eve.

[13] Referring to Eugenia von Mitzlaff's *Uppå korsets väg till lifsens krona* (Stockholm: Bonnier, 1879, 2 vols.), translated from the German: *Durch Kreuz zur Krone* (Halle an der Saale: Julius Fricke, 1864).

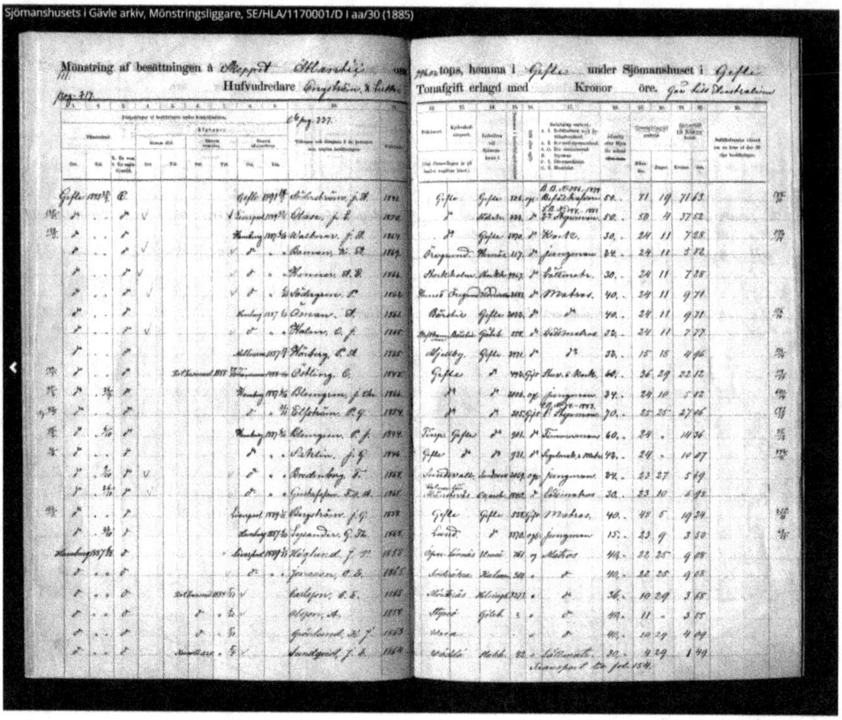

Fig. 10. *The enrolment register of the Atlantic in 1885.* Copy from a ledger in the Gefle mercantile marine office in the Swedish National Archives; public domain. Copy in private possession.

Now, as I write this down, Emy is participating, singing and playing the piano.

—Alas, how wonderful music is, and what a good company it is for us here on the sea!

But I, poor coward, have to content myself with just listening to all this!

Today, the weather has been sultry, oppressive and hot, but we are now under the tropics. However, after a tropical day follows a tropical evening and night with a starry sky, which in clarity and splendour is not surpassed even by a Nordic winter night.

Now we also have the moonlight; right now, it occurs to me that I am doing stupid things by sitting here and writing instead of going out on deck and spoon in the moonlight, which I now also intend to do!

The 27th of December, 1885

Our beloved Grandma's birthday! She is now 75 years old.[14] During the day, lots of thoughts have rushed home to her!

New Year's Eve, 1885

One year has now passed . . . a gracious year of the Lord! Looking back, many memories emerge, most of them very happy, and I realize that I have great reason to be very grateful to God. Yes, "nothing is lacking in all His good words." This is the language of the Bible that concludes the year in my birthday book, and it is especially appropriate at the end of the year to remember God's gracious promises and how He faithfully keeps them all. It should at the same time call for the new year to begin with hope and boldness!'

5. The Line-Crossing Rite of Passage

'New Year's Day, 1886

We have many joyful memories from this day!

Particularly worth mentioning is that today we cross the equator, *the Line*, as it is called in the language of sailors.

I would like to try to describe several larkish tricks that the crew were up to in connection with this passage. All that are not inscribed in the Neptune Order, i.e., those who have not previously crossed the line, must be properly "shaved" by Father Neptune! First mate *Elfström*[15] played the part of the god Neptune proper and was wearing the most priceless and fantastic costume of all time, masked and almost unrecognizable, of course.

The other dressed up persons that appeared were Neptune's wife, the doctor, and also policemen, all in very comical costumes. The very "shave" was done in the following way. Crew members were one by one led forward amidships by the police, with eyes closed or blindfolded,

---

[14] Born in 1810.
[15] Pehr Gustaf Elfström, born 1854, 31 years old, married; a highly valued steersman, navigator and much liked companion, better paid then the master. He signed off by death, less than five years later lost at sea.

where a tub full of water was standing. A number of questions, put by Neptune, had to be answered, and every time the candidates opened their mouths they were anointed with a brush of soot, putty, tar and all sorts of such cosmetics. Then all this was scraped off with a wooden knife, and finally the novices were overturned in the tub.

In the evening, a splendid donkey appeared with its driver, and all sorts of other practical jokes and performances took place, such as torchlight processions, etc., and all had fun until late in the evening.

Emy and I watched and laughed, but escaped the awful shave by paying a pledge, which consisted of a couple of whole punch cans for general treatment![16]

It often happens that we signal to ships that we meet. Today we met one named *Nanna*, which comes directly from Gefle like us, and is on her way to Australia.[17] The captain has his wife with him on the trip. The three of us were invited there for dinner, but were not particularly inclined to accept the invitation, but preferred to stay on board with the rest of the crew.

5 January, 1886

I have felt very happy today. The reason is that Dad, Emy, and I have had a very pleasant conversation on a subject, which should be closest to our hearts, namely about God and our relationship with Him. Alas, thank God, especially for having aroused in Father the fervent desire to belong to Him. Yes, I cannot describe how happy I felt when my father

---

[16] From the captain's storage of Swedish arrack punch, including wine merchant Zetterström's *Gefle-punsch*, brewed at the homeport.

[17] Like the *Atlantic*, the 42.1-metre-long *Nanna* with the signal JHRS (which after it was built in Germany in 1875 was to belong to various shipping-companies and change name a couple of times) was a three-masted barque, and on the same sailing routes in the international trade, but was not part of the Gefle fleet and never sailed in consort. They met from time to time though, and the captains made friends.

said that he firmly believed that a human being could not possibly feel happy without being a true Christian.

Among other things, we talked about the worldly pleasures: how a Christian cannot consider them pleasures, and how it is contrary to her conscience to participate in them, because they are contrary to the word of God. It is not possible to belong to God and the world at the same time. In that respect there is a choice to make, and it is a matter of choosing between life and death.

Alas, it is good and sweet to be able to confess that you belong to the Lord Jesus, and thereby to be able to remember His precious words: whoever confesses.'

[*sheet(s) missing in the journal between pp. 30 and 31 here*]

'However, concerning my own writ, I now close my notebook for the day with the firm decision and emphasis that no one will ever be allowed to read anything from it, not even Dad, as much as he ever wants, it will fail!

So happily, hand in hand
We go to the Phoenix-bird's land!
To the fairyland that shines
Of crystals and rubies!
We go to the Phoenix-bird's land!

The 12th of January, 1886

When I read through some of the pages here in the book, as I mentioned, I also found it very strange that, although I was travelling, I barely mentioned a word about the destination of our journey. Now much more than half the way is covered, and we have been over two months on our way. This time has seemed very short to us, which is why we have hardly yet begun to long for a port.

But on second thoughts, it will probably be enjoyable to reach Australia, which according to Dad's description if so wonderful. We'll probably have a lot of fun there. Dad knows of old several well-known families in Adelaide, so there are prospects of meeting and spending a lot of time with youngsters of our age (especially, it will probably be very interesting to get to know the so-called "Little Bumblebee", i.e.,

Miss Alma Hummel, about whom we have heard so much talk).[18] For that reason, we should be very diligent in studying English, but I must admit that I am quite negligent at times in this respect.'

6. Life in the Blue

'14 January, 1886

Yes, the days go by at an amble, the one almost like the other. Now and then it nevertheless happens that we get to see something special that attracts our attention.

Sometimes, it's a shark, some dolphins [Sw. *springare*, "springers"], a strange bird or fish, which we are very fascinated in watching. Today the crew caught a couple of very strange fish, in sailors' language called *bonetis*, but I really do not know what their real name is [referring to Skipjack tuna, oceanic bonito]. They are very large and thick, and beautifully glittering in many colours, have no scales but thick firm skin covering and are said to belong to the porpoise family. In the countries around here, they are eaten by the population, but cannot possibly be regarded as any delicacy, the meat is dark and woody and has an oily, cranberry-like taste.'

This is all about Euthynnus pelamys, named by Carolus Linnaeus in 1758, in 1829 renamed Katsuwonus vagans (Lesson), which is the streamlined Skipjack tuna without scales, also known as Oceanic Bonito, common in tropical waters, about three feet long, plump and shiny.[19]

---

[18] The Hummel family in Adelaide were great friends of captain Axel Söderström, and their daughter Alma was about the same age as Emy and Mia.

[19] Other fish they saw during this trip were porpoise, and they caught silvery-blue tenpounders, also called ladyfish and bonefish, a welcome gamefish too for the seafarers along with flying fish leaping out of the water and caught in nets or landing on deck. Fowl, domestic as well as migrant wild birds in their entourage, were also on the *carte du jour*. Although they passed many whaling boats, they did not see any whales until they reached the Northern Pacific in July, when they landed up in the middle of a flock.

'15 January, 1886

We are now just outside MONTEVIDEO.[20] Still excellent nice weather, though too little wind, for our ship to really live up to its name for being a fast sailer.'

They are now in the South Atlantic Ocean, nearing 34° 53' 1" S; 56° 10' 55" W, sneaking down the South American coast before turning left for the southern extremity of Africa when they catch the favourable eastward winds and currents, which advantageously gear sailing ships past Africa straight to Australia, and then avoiding the circular motions and whirls up in the Indian Ocean. Due to the Coriolis effect, there are westerly trade winds in the Southern Ocean, and sea-currents are primarily wind driven and therefore mirror the atmospheric currents. These conditions have been used by captains of sailing ships for centuries for travels between the continents. A long and chilly distance to travel in winter, in any case, over ten thousand kilometres, but, luckily enough, now it was summer in the Southern passage with an average temperature of some twenty degrees Celsius, except when the proximity of the Antarctic Ocean made itself felt. Montevideo is the capital of Uruguay, which had become an independent state in 1828. The city was getting modern at this time, as the first telephone lines were installed there in 1882, and electric streetlights took the place of the gas-operated ones in 1886 just as the Atlantic passed.

'Dad, Emy, and I had a confidential conversation today about our dear dead mother. We seldom talk about her, but I often wish so fervently that she had lived a few more years, so that we could remember her better.

17 January, 1886

Often during the journey, the sad thought has crossed my mind that we will probably never see Grandma again here on earth. Last night I dreamt that she was very ill. A great deep consolation lies in the

---

[20] They were close enough to see the lights of the Faro Punta Brava (erected in 1876) and Faro de Cabo Polonio (1881–) over fifteen nautical miles away. The waters bordering on the discharge of the Río de la Plata, the 'Silver River', were familiar to the crew who knew the entrance to Buenos Aires on the opposite shore of the river.

certainty that once in heaven I will see my loved ones again, those who have gone before us. Yes, I often remember Grandma's parting words, and I want to try to often submissively and confidently repeat the words "God's will be done."

A stanza from a little song reads as follows:

> Only the thought of home makes the pilgrim strong,
> He does not dare to rest on uncertain ground.[21]

Alas! Would that I more and more often, and at any time and moment, could have my mind fixed on Jesus, and on the home up there, to which I have access through Him. If I could leave aside all my own aspirations and trust Jesus completely.

As for getting to heaven once, I'm sure that through Jesus I will happily overcome, and be saved until then. He probably does not want to let me go and let me get lost, for that He loves me too much, and His sacrifice for me is too expensive to be wasted. May He, through His Holy Spirit, give me an ever firmer will to belong to Him, and an ever more zealous, heartfelt desire and longing, to completely sacrifice myself for Him and for His service. Alas, I wish I had to forsake something for Him, who has forsaken so much for me! Alas, that I might do something for Him to prove the love which He Himself kindled in my heart! But look, I now have again that same self-will (to want to be and do something myself) for which I sacrifice day by day, I often have to endure difficult battles, in which I usually sustain a defeat, as a result of "going ahead in my own power". It is also often the case that I do not care but am indifferent and let evil prevail. God help me, poor child!—It is required that I seriously pray for the right childlike mind, which will make me understand how little I can do from myself, and about humbleness above all. In this I have to follow the example of Jesus, and receive from Him power, that "His power may be perfected in my weakness."[22]

---

[21] Once more, Mia is quoting a familiar song worded in Swedish by the poet and hymnist Lina Sandell (1832–1903), this one originally inspired by Hebrews 11:13-14, on 'strangers and exiles on the earth [...] seeking a homeland. If they had been thinking of that land from which they had gone out, they would have had opportunity to return.'

[22] The Biblical allusions in this paragraph refer to Paul, 2 Cor. 12.

7. The Albatross

'25 January, 1886

Today is a cold and awfully rainy day, reminiscent of an October day at home. Yet it is rather homely to hear the rain pattering against the windows. We are quite far south now, so that it has become very chilly, at least it feels so as a result of the rapid change from tropical heat to only about +9 to 10 C°. But we Northerners are hardened!

A couple of days last week we had full storm; then the ship meandered and careened terribly. Emy and I withdraw and were lying but were not in the least seasick, but lay and read, ate well, talked, laughed and sang, so that it was heard on deck; all that is a good sign indicating that we are gradually becoming capable seamen!

27 January, 1886

Today we caught the first albatrosses.

Alas, how stately they look with their white, delightfully beautiful feathers, and immensely large wings. One of those caught today is 10 feet between the wingtips. The longest wingspans of the genera are about 14 to 15 feet, I have been told.'

INTERMEZZO: To the deep-sea travellers these majestic companions were both a marvel and a comestible, and the glaring contrast between the soaring white giant hovering in the sky and the crashed game in its last throes suspended on deck, inspired deep thoughts among many of those who risked life and limb on sea and were used to perils out of the blue and changing fortunes. The crews were well aware of the symbolic nature of the bird, as were those who had majored in literature at the Själander Girl's School. Captain, mate, and others of the older experienced hands were familiar with some verses from the world-famous "The Rime of the Ancyent Marinere' (1798) by Samuel Taylor Coleridge:

> *At length did cross an Albatross,*
> *Through the fog it came;*
> *As if it had been a Christian soul,*
> *We hailed it in God's name.*

. . . .
And a good south wind sprung up behind;
The Albatross did follow,
And every day, for food or play,
Came to the mariner's hollo!

In mist or cloud, on mast or shroud,
It perched for vespers nine;
Whiles all the night, through fog-smoke white,
Glimmered the white Moon-shine.

God save thee, ancient Mariner!
From the fiends, that plague thee thus!—
Why look'st thou so?—With my cross-bow
I shot the ALBATROSS.

A colleague among seamen, Herman Melville, in his book on the whaling ship PEQUOD ('Moby-Dick; or, The Whale' (1851)) had pondered upon 'those clouds of spiritual wonderment and pale dread, in which that white phantom sails in all imagination.'

Even more recently, Charles Baudelaire in the second edition of 'Les Fleurs du mal' had added to the metaphor of life and death, guilt and fate, and the dizzy height and pitfall of all aims in L'Albatros' (1861) in its elegy to:

those vast birds of the sea
That follow unwearied the voyage through, Flying in slow and elegant circles above the mast.
. . . .
How droll is the poor floundering creature, how limp and weak — He, but a moment past so lordly, flying in state!

'We are approaching the Indian Ocean boundary in the Southern Ocean and it is cold, *huh* so cold! We have a good fresh breeze with ample seaway and we are getting on fine. We can be at the CAPE OF GOOD HOPE at the end of the week, and from there during normal circumstances with average speed it takes usually … [*sheet(s) missing in journal betwixt pp. 38-39; i.e., the passage of the Cape and the crossing of the Indian Ocean*]… journey.

Yes, the Lord is gracious, and it is lucky to be able to rely upon Him. And always, for time and eternity, I must enclose myself in His gracious protection, for so His word reads: "with everlasting kindness I will have mercy on you".'[23]

8. To Australia: *Adelaide*

On Monday 1 March the SOUTH AUSTRALIAN ADVERTISER *announced under* CAPE BORDA SHIPPING, *with a weather report: 'February 27 at 4.30 p.m.—Swedish barque, signals H.N.J.B. from Gulf of Bothnia passing inward. Wind—S.S.E., fresh; sea moderate.'*

The journal also listed under IMPORTS: *'*ATLANTIC, *from Gefle, 25.212 pieces of deals and battens' [which was part of its regular cargo on the route: timber, planks, deals and battens].*

The same day there were under SHIPPING INTELLIGENCE *the following communications in the* SOUTH AUSTRALIAN REGISTER *(Adelaide, South Australia): 'Port Adelaide Arrived: Sunday, February 28,* ATLANTIC, *barque, 996 tons, J. A. Soderstrom, master, from Gefle November 7, Elsinore November 15, C. Gunnersen, agent.'*

In the REGISTER, *there was also a very informative article about the ship and its voyage under* MISCELLANEOUS:

'The ATLANTIC, *barque, arrived again on Sunday in her usual excellent order. Captain Soderstrom reports leaving Skutskar on November 7, and on reaching Elsinore was there detained windbound for eight days. She took final departure on November 15, and had fine weather across the North Sea and down the Channel, clearing the Lizard* [Lizard Point is the southernmost point of the British mainland] *on December 6, and having a westerly gale for a day or two, and then easterly winds for three or four days. On January 1 crossed the Line, and on the 31st was in the longitude of the Cape. During the passage of the Southern Ocean she reached south to 46°, but saw no ice.*

When in 50° E. 46° S. fell in with a vessel looking like an iron barque, which had carried away her jib boom [the spar run out forward as an extension of the bowsprit], *and was rigging out a spare spar, but she declined to signal her name. Two other vessels were spoken bound from the*

---

[23] Isaiah 54:8.

Baltic to Melbourne, but no other incident of importance occurred during the voyage. The ATLANTIC *came to anchor at the Bell Buoy early on Sunday, and on Monday Captain Soderstrom would land to seek orders for destination of cargo.'*

The diary continues:

'ADELAIDE 28 FEBRUARY, 1886 [SUNDAY]

To be able to cast anchor in this earthly haven, Oh, what a good and pleasant thing it has been! How will it not also be, that after all the storms of mortal life, to end up in heaven!

Exactly on my name day we arrived, Mary's Day on 28 February. What a suitable way to celebrate it! It was further commemorated to such an extent that my name was set on fire and a "Hurrah for Miss Mary" was called for by the crew and cheers resounded here on board (Figure 11).

—Dad is now ashore to pick up the mail. It is with almost feverish longing and anxiety that we await him back with the post.

1 MARCH, 1886

Not before today we received the correspondence. No letters were handed out yesterday, because it was Sunday. Here you immediately meet with English manners and customs.

The most welcome letter, of course, was that from Grandma. It was with mixed tears of joy and sorrow that I read it. Oh, it feels terribly painful to think about how far apart we are from her, and how long it may take before we see her again.

The news in her letter about Tante Klintberg's and Mrs Öster's deaths were startling.[24] And furthermore that she therefore has to move, poor grandmother, so awkward and worrying this becomes for her.

---

[24] Clara Klintberg (1822–1885) was the daughter of the municipal medical officer in Gefle, Carl Nordbladh (known for his fear of being buried in suspended animation like many others were those days), and married to the Headmaster of Gefle Upper Secondary School, Anders Gustaf Klintberg (1816–1884). Mrs Klintberg was related to Emy's and Mia's so-called 'Grandmother' Charlotta Winroth, née Leufvenius, and

Fig. 11. *The 1885 crew of the Atlantic pictured in Melbourne in March 1886.* Courtesy of the National Maritime Museum, Stockholm. Fo24248. The crew of the *Atlantic* photographed upon their arrival to Melbourne in March 1886. The personnel living in the stern castle (upper right) have had opportunity to compose their features and dress up to the occasion, whereas the rest of the staff is still in their regular working-clothes and worn after the long trip. *Standing, from the left*: Carpenter Per Johan Blomgren, b. 1844, 32 years, married; ordinary seaman C. J. Holm, b. 1865, 21 years; deckhand Knut Axel Boman, b. 1869, 17 years; ordinary seaman F. O. A. Gustafsson, b. 1865, 21 years; ordinary seaman Per August Hörberg, b. 1865, 21 years; deckhand Gustaf Theodor Lysander, b. 1868, 18 years; ordinary seaman A. E. Hanner, b. 1866, 20 years; able seaman Anders Öman, b. 1861, 24 years; deckhand Johan Christian Blomgren, b. 1866, 20 years; the steward and chef Carl Östling, b. 1848, 37 years, married; second mate Johan Leonard Clase, b. 1850, 35 years; Emmy Söderström, b. 1868, 17 years; Maria Söderström, b. 1867, 18 years; first mate Pehr Gustaf Elfström, b. 1854, 31 years, married; and captain J. A. Söderström, b. 1841, 45 years, widower. *Sitting, from the left*: Able seaman and sailmaker Johan Gustaf Bergström, b. 1838, 48 years; able seaman Petrus Södergren, b. 1862, 23 years; ordinary seaman F. Bredenberg, f. 1868, 18 years; the cook Johan Axel Waltner, b. 1864, 22 years; and the sailmaker and seaman Johan Gustaf Sahlin, b. 1846, 40 years, married. Finally, the third female hand, the helpful ship's hound Harriet, a St. John's water dog, an extinct breed. The monthly salary of the employees varied widely, from 15 kr for deckhand Lysander to 70 kr for chief officer Elfström, but food and lodging were included with the round-the-clock attendance.

64   Ocean Bound Women

There were eight letters each for Emy and me, yet not everyone who promised to write had kept his word. But the dear, dear friends, Lully, Hanna and Hedvig have not forgotten us, nor have Maria W. and Selma F.

We will now be busy answering all these letters; but this is quite a pleasure.

In the afternoon we were ashore on visits.

Dad is acquainted with several families here. So, accordingly we went out shopping, and received many beautiful gifts from Dad; including a very cute hat each. There are magnificent stores here. One of them extends along an entire block with five to six storeys high houses and has *300 shop assistants*.[25]

Gas street lightning had been introduced in PORT ADELAIDE *in 1881 and the harbour facilities were expanding as the town grew fast: a number of fashionable commercial and institutional buildings were erected in the 1880s (some of them still partly standing) in this seaside city in the colonial South Australia State in the Commonwealth of Australia, under the British crown (as it still is today). — Once so famous* JOHN MARTIN'S *(1866–1998), flagship departmental store in the* RUNDLE STEEET *business precinct, Adelaide, at that time reached its first heyday, along with J.* CRAVEN'S, *which opened in 1886.* FOY'S *were later to follow. John Martin's original business companion in* PETERS & MARTIN, *Otto Peters, was lost in*

---

the couple had cared for their biological mother and her offspring until her early death, as is evidenced by letters. Clara obviously remained a *Tante* or 'aunt' to her small children, as we can see. Wily gossip later indicated that the concerned headmaster was in reality their paternal grandfather, whereas family legend had it that their mother was an illegitimate child of the regularly touring King of Sweden—an embellishing story they shared with thousands of other families with unknown paternities, fatherless children, and forlorn women!

[25] Maria Söderström's awed description of big Australian cities was corroborated almost in the same words and spirit by the French artist Paul Gauguin on passage through Melbourne and Sydney to the South Seas a few years later, stunned by the high-rise blocks, steam trams, and the familiar London cabs, besides considerations of low living costs and easily earned money in 1891.

the wreck of the barquentine steamship GOTHENBURG, *which went down in a tropical cyclone at the Great Barrier Reef a decade before, in 1875.*

'2 MARCH, 1886

We stayed ashore last night in a hotel run by a family named Ford.²⁶ We became really good friends with their daughter Bessie, she was really nice. Everyone shows us so much kindness, especially a Swedish family named Hummel; we were invited there for dinner and had quite a nice time with their daughter Alma. Oh, what a little mite she is! She has urged us to stay with her the entire time we are here in Adelaide. In the evening we returned aboard the *Atlantic* and had a whole boatload of strangers with us. They stayed until close upon twelve o'clock at night. I was not really happy but felt somewhat downcast and anxious. It's so weird to come from the simple and still life on the sea to all this hullabaloo.

3 MARCH, 1886

Morning. I have been busy writing a long, long letter to my beloved grandmother. Today I thank God I had time to pray and read God's word. I read something from Moody's talks; they are so beautiful and uplifting. Incessantly urging to turn wholeheartedly to the Lord, believe and be saved!²⁷

---

²⁶ Ford's Hotel, also known in Clare, SA. There were dozens of hotels in the port and numerous in the town.
²⁷ Dwight L. Moody (1837–1899) was a North American revivalist whose career started as a shop assistant. In the mid-1850s, after a severe religious crisis, he converted to evangelical Christianity and worked energetically in the Chicago congregational mission. Huge crowds gathered for his and his aide the singer Ira Sankey's revivalist meetings, and Moody became one of the main figures in the great American revivalist movements of the 1870s. Through travels to England and Scotland, he helped spreading this revivalism to Europe as well. Between 1875 and 1876 several of his lectures and sermons were published in Swedish translations. Apparently, there was a copy on the *Atlantic*.

Afternoon. We went ashore again and got Alma out with us to see the city. We went around hither and thither to the sights, such as the Botanic Garden, which is magnificently beautiful with all kinds of strange Australian plants. Many of those that we see at home in miniature as potted plants, such as *india-rubber tree, cypress, cactus*, etc., grow here as tall trees.[28] Furthermore, we visited a picture gallery, a museum, etc.

Another Swedish family named *Lindström*, I should also tell a few words about, and in particular on the daughter in the house, nicknamed *the grass-widow*, who should deserve a chapter of her own. And, of course, I could have very interesting things to write about her, because she really open-heartedly told me her whole life story, with love adventures and all. It was a pity that her younger sister, Carrie, the famous music teacher, is not at home but away at the moment, so that we do not get to know her.

—An engineer Dan Lundberg from Gefle, who a few years ago was a passenger out here with father's ship, we met: we had dinner with him today.[29] He is now married to an Englishwoman; we were invited to them but did not have time to go—for we must already now begin to prepare for the departure from here, since it has wholly unexpectedly been decided that we may unload our cargo in Melbourne.[30] Thus we must then part from all our newly acquired acquaintances and leave!

---

[28] The Adelaide Botanic Garden was opened to the public in 1857, and in 1886 it comprised the vast Botanic Park, the Adelaide Zoo, and also a Palm House since 1877.

[29] The *Lindstrom* and *Lundberg* families were from Gästrikland and Upland in Sweden, as the party was, and acquaintances of the captain, which all had settled down in Australia.

[30] The remainder of the timber, wood products, woodware, and building material in the large commercial cargo taken in at *Skutskär* in November 1885, and brought to Adelaide, would be sold at Melbourne in March 1886. The demand for woodware had earlier been booming in Australia, but the export market had become unpredictable and competition was rocketing. In every harbour, shiploads were bargained and auctioned.

On the 4th of March we shall sail. It is calculated to about a week's voyage.[31] Hope that all goes well! Yes, we want to take God with us on the journey.

We have seen in the newspapers from home how unfortunate it has been for several ships that passed the North Sea only a couple of weeks later than us. Yes, it may well be that impending danger has often been near to us, too, although the Lord graciously helped us and made everything go well.

7 March, 1886

It's Sunday today; it is stormy both outside and inside. That is, it's awful weather, raw and chilly, and maybe that's why I also feel that cold and gloomy inside, bleak as the atmosphere. There are times when I feel so dreadfully uneasy and disquieted. Today it is so for no particular reason.

My heart is like the sea with all its ceaseless veers

Would that I could also truthfully add the following to the verse:

Yet many a precious pearl it in its bosom bears

Now I want to pray to the Lord God that He will reveal Himself to me again with His Holy Spirit, dispel the darkness and give me His peace. This peace has probably not departed from me, it still feels that way. No, His peace is eternal and unchanging as He Himself. This

---

[31] The basis of this time calculation might seem perplexing to anyone who is unfamiliar with the vastness of the continent Down Under and tall-ship navigation. However, the distance from Port of Adelaide to Sandridge Pier (Port of Melbourne) was 556 nautical miles, and time at sea with an average speed of five knots would ideally be 4.6 days and nights at straight stretch if it went like a clockwork—disregarding rocks and reefs and towing at two difficult harbour entrances. The seasoned captain had done this before, knew the obstacles, and was no hoper. The sisters had been sailing with him before, as their mother once had done in the past, and they would confidently do it all over again.

thought should be encouraging. Oh, that I could therefore "be still before the Lord and with a steadfast heart trust in Him."[32]

We have good wind, and it goes on fast, so we will soon reach our destination. These days we are busy reading the newspapers, which we received from home, and reading over and over again the dear letters, and enjoy our supply of fresh fruit, the most delicious grapes, pears, and oranges etc. The grapes in Adelaide cost about 10 öre per pound [lb].[33]

We wrote to Auntie and asked her to let Grandma move in with her, it would be so nice in every way.[34]

If I think back on the times that have passed, especially last year (I feel particularly bent on reflecting right now), I find that it is a thousand times more pleasant this year than last year! Alas, then I also longed a lot for the last of February, when I would come home over cousin Elma's wedding,[35] and that was very nice, especially because I then got to meet my beloved grandmother there; but a few days afterwards I had to return to the eerie Ofvanåker.[36] The time I stayed

---

[32] Allusions to the *Psalms*.

[33] Corresponding to about 7 kr in 2021. 10 *öre* was the common Swedish letter rate since 1885, with a *ten-öre*-stamp accordingly issued that year featuring King Oscar II's portrait (the first Swedish stamp with a picture). So, the postage stamps on the letters were worth about a pound of fine grapes in South Australia then. In Sweden, poor people could get a night on a dosshouse for that amount of money.

[34] Referring to their paternal aunt Catharina Charlotta Zellinger, née Söderström (b. 1833), widow since the death of her husband Pehr Zellinger at St. Helena in 1864 (the parents of their cherished older friend Ida who was born in 1861). Their other surviving aunt, the elder Johanna Katarina (b. 1825), was the mother of Johan Leonard Clase (b. 1850), Second Mate of the *Atlantic*, who kept in touch with his parents on his own. The two aunts Charlotta and Johanna lived until their nineties and their adoptive 'Grandma' to 83, whereas men in these seafaring families were more transient and transitory.

[35] Elma Charlotta Leontina Andersdotter Clase, Aunt Johanna's daughter, born in 1852 and married to cabinet-maker Johan Erik Forsberg (b. 1847).

[36] *Ovanåker* is a small village in Gävleborg County, Sweden, the original vicariate of the renowned *Celsius* family, where Mia was temporarily accommodated in a hospitable Samuelsson household between the school terms on winter holiday, feeling more homeless than ever. Like her sister and other orphaned teenage girls, she occasionally acted as a maid or nursemaid. The only place where she would feel quite at home in her teens was on the sea, because then she was with her sister in reunion with their father, after over a decade on land without parents.

there was probably good for me, although it was boring many times. Still, I will always remember my sweet little aunt Sam., and the cute little kids there. Concerning "the old man" I cannot help that I have no fond memories of him. But phew! How mean I am! Despite all his peculiarities, he was probably nevertheless a good, kind, and sincere Christian, basically. Do I dare to doubt this?'

9. *Melbourne*

'SANDRIDGE PIER is very close to the railway station, from which trains depart every half-hour up to the town.[37] Hitherto we have made a trip up there each day.

We have also made several pleasurable acquaintances. Every day we are accompanied by a Captain Gädda [meaning 'Pike'],[38] a good friend of Dad. He always brings *"yum-yum* for the little girls".

In addition, we are also very much in the company of a Captain Sätterlund,[39] whom we first met in Elsinore, and also signalled to a couple of times on the sea during passage. We like him a lot.'

Birger Gotthard SÄTTERLUND, *the master of the fabulous barque* AURORA, *was born in 1852 in Haparanda, Lapland. He usually brought his wife Hilma on his trips, and for that reason the captain had to act as midwife when their daughter* AURORA *was born on the open sea, named after the ship. During a trip in 1886, a speed record was beat on a voyage between Newcastle, Australia, and Honolulu, Hawaii. In 39 days, they sailed 6,075 nautical miles (11,250 km), making an average of 156 miles per day. The average speed was 6½ knots (top speed 10 knots as for the* ATLANTIC), *and the voyage, which mostly went with all the sails set, made the* AURORA *and Captain* SÄTTERLUND *famous all over the world at the time.*

---

[37] At the pier, ships were onloaded and unloaded by goods wagons, and shiploads transported by railway, and there was also passenger traffic by rail uptown.
[38] Gädda, Karl Henrik, born in 1839 in Skredsvik, *Göteborgs och Bohus län*, a sea captain from Gothenburg, the most important Swedish harbour city in the international trade. *Gädda* is on old Swedish family name, known since the fourteenth century.
[39] His fine SEXTANT is still preserved at the Norrbottens Museum in Luleå.

'Captain Grubbström,[40] who has his wife with him—of the ship *Nanna*, which we met at sea on New Year's Day!—we have also seen again.

Furthermore, a young Norwegian, Mr Römcke,[41] whom we got to know two years ago in Drammen [Norway], we have renewed our acquaintance with.[42] He invited us to dinner the other day at the hotel *Maison dorée*,[43] in company with some others, including a young Gothenburg captain named Ljungberg, whom we liked quite a lot![44] Mr Römcke is engaged to a Scotswoman, and is to get married next month. We'll probably be invited to the wedding, if we're still here then!'

This young Norwegian, Otto Römcke, had come to Australia as an envoy to study the market and decided to settle down in Melbourne permanently. There he established himself as an agent and importer of timber—see here the connection with the commercial cargo the Atlantic carried and offered

---

[40] Isak Petter Grubbström, born 4 March 1837 in Bygdeå Dalkarlså, master of the barque *Nanna*.

[41] Mr Römcke was finally appointed Honorary Consul-General for Norway in 1906, which was announced in the *Commonwealth of Australia Gazette*, No. 43, Saturday, 11 August 1906, on the front page. His life story and doings, along with that of other Scandinavians, are outlined in Jens Sorensen Lyng's *The Scandinavians in Australia, New Zealand, and the Western Pacific* (Melbourne University Press, 1933), where Römcke is mentioned on pp. 81, 94f.

[42] This corroborates the family saying that one or two of the sisters were sailing with the *Atlantic* in 1884 and thus were fellow-travellers on the former trip.

[43] Fasoli's *Maison Dorée* ('the golden house') was to remain a fashionable French-styled *Café* and meeting place in the 1890s, inspired by the *Café de Paris*, which also opened in Melbourne.

[44] Captain J. E. Ljungberg!—i.e., this was the first meeting with Mia's husband to be, the master mariner of the sailing ship *Gurli* of Gothenburg, a barque of 721 tons owned by George Douglas Kennedy. It was built at the old shipbuilding yard in Gothenburg in 1879. The similar and later famous *Sigyn* with three masts of pitch pine was also built there, today a museum ship in floating dock on display in Turku, Finland. The *Gurli* was before long to return to Europe with a shipload of wheat, but when the couple at last were married their firstborn child was named Gurli after the ship. Gurli's son Yngve Herrström, second cousin of the present author, was the original inspiration behind the writing of this work.

for sale in this seaport, which was partly a Norwegian–Australian trade agreement!

'Yesterday was Sunday, and some people came to see us here on board. They were, among others, a Mrs Zickerman, from Stockholm, who has a small hotel here in Melbourne. She seemed kind and nice.[45] Furthermore, Mr Römcke, and a Mr Carl Grundén, formerly living in Gefle and old acquaintance of father.[46]

There is a Swedish-Norwegian church here. I have rejoiced so much about this and have been looking forward to go there to hear the word of God, but sadly I learnt the other day that the pastor is gone away and will not return in a month.[47]

We've been to the theatre once: we saw the *Boccaccio*. Oh dear! Damnation![48]

---

[45] Ida Cornelia Fredrika Zickerman, née Sjölander Stenbeck (b. 1859), of the *Rose and Crown Hotel*, Melbourne. She had married Wilhelm A. Zickerman in Massachusetts, USA, and arrived in Melbourne on 15 May 1884, by the *SS Cheviot*. This Australia-based schooner-rigged steamer was built in England 1870 and was wrecked in high seas at Point Nepean, Victoria, in 1887, with the loss of thirty-five lives. It is noted in contemporary public notices that Mrs Zickerman was unluckily fined in 1884 for (allegedly) 'detracting from the strength of Hennessey brandy' at the hotel (*The Geelong Advertiser* (VIC), Tuesday, August 12, 1884.) According to the Hallengren family archives, she was to leave Australia later on and go to Sweden. A street in Skövde, Sweden, is named for a famous textile designer in the family, *Zickerman's Street*.

[46] The first major inflow of Sweden-born immigrants to Victoria occurred during the gold rush of the early 1850s. Based in the goldfields of Ballarat, Bendigo and McIvor, many Sweden-born immigrants eventually sought work on the land as farmers, shepherds, and wine yard workers and vintners, whereas others settled in Melbourne as craftsmen, builders, shop assistants, businessmen, and service staff. The Swedish Club was formed in Melbourne in 1887.

[47] Pastor Carlsen of the Scandinavian Church, later the Swedish Church in Melbourne (today in Toorak). Sweden and Norway formed a union, the United Kingdoms of Sweden and Norway, in the years 1814–1905.

[48] They were to the Alexandra Theatre (later renamed HER MAJESTY'S THEATRE), and later attended premieres at the Princess' Theatre in Melbourne, both buildings finely restored, preserved and still standing, as is the Royal Exhibition Building and other places they visited. They were invariably impressed, but concerning plays and

23 March 1886

Daddy's and Emy's nameday: Axel [/Axelina]! They've received great many bunches of flowers. We've been to dinner today aboard the *Nanna*. It was quite nice.

24 March 1886

Dad caught a cold last night, so he's sick today. Hope it's not dangerous!

Today is the first day since we came here that we are in peace and quiet for ourselves. We've been busy writing letters. We have received even more letters from home now, another from Lully, and from Aunt Sam., and Aunt Klara.

On 1 April, 1886 [Thursday]

You usually have a foretaste of spring on the first of April! That is, at home in dear old Sweden! What I think is, that the people who have moved out here would miss that pleasant shift between different seasons, which at home is always so refreshing! Here it is practically as good as eternal summer!

Today, thoughts are constantly wandering home. I wonder if anyone there is also thinking of me right now. Oh yes, of course my little grandmother does so! I feel it!

Last week we had Carrie Lindström—from Adelaide—here with us for a few days. She was here in transit from Sydney, and left last Saturday with a magnificent French steamer. We followed her on board there on departure, and Dad was so liberal and offered champagne then! From Carrie's sister, Mrs Ramenius,[49] we have received small, nice gifts as souvenirs. And Alma H. has sent me a little flower vase, which she has painted herself.

---

performances they were divided. *Boccaccio, or the Prince of Palermo* was a comic opera in three acts with music by Franz von Suppé and text by Zell and Genée, which was first produced at the Carl Theatre, Vienna, in 1879. It was a funny and indecently light operetta with a theme from the renaissance and ran on stage through the 1890s with Australian actors and producers. It was first performed in Melbourne at the Prince of Wales Opera House in 1882.

[49] On the 6 of January 1868, the renowned music mistress Carrie Lindström's younger sister Ottoline Louise Lindström, second daughter of Carl and Christine Maria Lindström of Gothenburg, had married Carl Johan Ramenius, esq., of Dalarna,

Last Sunday we were to church. The Norwegian priest is now back but only on a short visit. It was very nice to hear the word of God, but conversely, I was rather distracted. The sermon was certainly quite good, but I probably wished it was in Swedish instead of in Norwegian. When we came from the church we were invited to dinner by Captain Gädda. We had an excellent evening. —

On Tuesday, Mrs Zickerman invited us to dinner and for a ride in a [four-wheeled horse-drawn] cab [brougham or landau] around the city and adjacent areas. We were to a large museum and an art gallery, much more splendid and impressive than the ones in Adelaide.

We furthermore walked in the park at the EXHIBITION BUILDING, and visited all kinds of sights nearby, including an Aquarium. —

Yesterday we were to the ROYAL BOTANIC GARDENS. It is indescribably beautiful! The governor's palace is located nearby, and it is also really magnificent. —

Today we have once again been fairly quiet and at rest. It is very hot today and we are tired after yesterday's walk. Nevertheless, uncle Gädda is now here to persuade us to take a walk at dusk, and he has good persuasive ability, so who can resist?

ON 3 APRIL 1886

Today mail has arrived from home, but there were no letters for us! We wait and long for Aunt's and Grandma's second letter. I cannot understand why we have not yet received them. May nothing sad have happened to our loved ones back home! Through Leonard we have heard that little Svea has been very ill.[50] Poor little thing!

Sweden, at St. Paul's Church, Adelaide—as announced in the *South Australian Chronicle and Weekly Mail* (Adelaide, SA) of 11 Jan. 1868. Carl Lindstrom became a member of the South Australia Chamber of Commerce.

[50] The second mate on the *Atlantic,* Johan *Leonard* Clase (b. 1850), who preferred to be referred to by the name of *da Vinci*. The little girl was *Svea* Katarina Kock, born in 1877, the firstborn of Aunt Charlotta Zellinger's daughter Evelina's children, thus a close relation like the second mate.

Today we have been invited to go by a young Norwegian captain named Andersen.[51] We went to the Zoo, and there was a lot to see. It was a large collection of both Australian and other strange animals. There was also music, and a large crowd of people in elegant toilets. Among others, we saw the governor[52] and his family, they went in four-wheelers and had outriders. Oh, how the ladies here stoke up and excel in their toilets! Silk, velvet and diamonds—and this is on the streets, on walks and in the shops, where they appear in full gala dress!

On 8 April 1886

Captain Gädda put to sea. Dad, we, and Captain Zetterlund[sic!] went out early in the morning to the roadstead to say a final goodbye to him. It will probably be empty afterwards, he is leaving a blank behind him since he was so nice and kind, and we always had a nice time in his company.

It has now been decided where we can go from here next, and that is to a place on the West Coast of North America, PORT TOWNSEND at PUGET SOUND.[53] We go ballasted there to load wood and timber

---

[51] Captain Theodor Andersen, whose brig *Oberlin* had run aground in Tasmania in 1875. One of the 330 *Andersen* immigrants listed by the Australian government 1848–1912, arriving with the merchant fleet. https://data.gov.au/data/storage/f/2013-05-12T195404/tmpsLxwENImmigration-1848-1912.txt

[52] The Governor of Victoria since 1880, Henry Brougham Loch, 1 Baron Loch; British Colonial Official born in Scotland who had earlier been imprisoned in China, a remarkable man observed elsewhere in the diary.

[53] Port Townsend (Jefferson County, Washington) at Puget Sound. Puget Sound is a bay with many islands and interconnected marine waterways and basins between the Cascade Mountains and the Coast Range. The bay is connected by the Juan-de-Fuca Strait to the Pacific Ocean and nowadays has significant traffic to and from the ports of Seattle and Tacoma. Seattle was founded in the early 1850s and in 1870 had little more than a thousand inhabitants but in 1900 over eighty thousand. In Tacoma, as in Seattle, there are many Swedish descendants; a Swede started a sawmill there in 1852. Tacoma became a town in 1874, and in 1880 there were just over a thousand inhabitants. In 1886 there were still an abundance of American Indian natives in the mountainous forest areas of Jefferson County, including the Chimakum, the Hoh, Clallam, and Twana tribes, called 'woodsmen' by palefaces.

for Melbourne, so we will get back here again. Through a couple of American captains, we have heard that it is a nice and agreeable place with a pleasant climate, similar to that of Scandinavia with its four seasons. We will arrive there in the fairest season, at midsummer time; it can be estimated about 2 ½ months of travel to go thither.

10 April, 1886

Today we have again been 'partying' all day. We have been invited to go for a ride twice, by two different people. Early in the afternoon we were invited by an English merchant named Caulson.[54] We travelled through several of the so-called *suburbs*, up the YARRA RIVER.[55] It is magnificently scenic. We had already seen some of the river before, as we have a few times in the evenings travelled there by *tramway* to enjoy country air, but we have never been as high up as now.

Later in the evening we were invited to an extremely generous dinner by Captain Swan.[56] He is a Finn but has been in America for many years and have sailed with ships from San Francisco. He is exceptionally nice and kind, with masculine looks and very handsome. So, then we were to the theatre again and saw RIP VAN WINKLE—and that play was really the best I've ever seen on the stage; with exquisite dazzling music, magnificent scenery, etc.'[57]

---

[54] Probably *Coulson*. Unidentified in the family papers. Might be Arthur Coulson, married to Elizabeth Parkin, Melbourne citizens; or Charles Coulson (1865–1940), who became a famous football player, or Albert Coulson, who graduated from the University of Melbourne, or… that name was well established in Melbourne.
[55] The first European explorers to sail up this river in 1803 were headed by the surveyor Charles Grimes, later followed by John Batman, who observed that a site on the northern bank of the river was suitable for the Port Phillip town project. Sea captains heading for Melbourne in the nineteenth century wended their way towards Port Phillip, the name on the nautical chart.
[56] Originally spelt Swahn, a Finland-Swede.
[57] Performed at the Prince of Wales Opera House, in Melbourne, VIC, since 1884, and then at the Theatre Royal. The eminent adaptations of Washington Irving's short stories 'Rip Van Winkle' (1819) and 'The Legend of Sleepy Hollow' (1820) for the stage had been successes in London and New York already in the mid-1860s. A main

Fig. 12. *Maria Söderström, the diarist, in Australia in 1886.* The Maria L. Hallengren Collection.

Fig. 13. *Emmy Söderström on the voyage.* The Maria L. Hallengren Collection.

10. Ball with the Imperial Russian Navy

'22 April 1886 [Thursday]

We do not have many days more to stay here in Melbourne now. Maybe we will sail already on Saturday. Consequently, we will probably celebrate our Easter on the sea, with exception for Good Friday, to-morrow. I'm very happy to be able to go to church then. Pastor Carlssen [the Swedish-Norwegian minister for Melbourne] is here over Easter.

On this return trip across the Pacific Ocean, from southwest to northeast, to North America and back here to Melbourne,[58] we intend to bring with us as a travel companion, a little kind and nice girl, Mary Grundén. She is the daughter of Carl Grundén, about whom I have probably already mentioned a few particulars before in my notes. She cannot speak a word of Swedish, and it is for teaching us English that she accompanies us on the voyage, that is the idea. And now it is my serious intention to be really diligent with my English lessons.

—A long time has passed now since the last time I wrote in my diary, and much has happened since then. Among other things, we have been at a wedding, and to a grand ball.

Mr Römcke married on the 13th of this month. It was a church marriage with a minister of religion officiating at the ceremony. The event did not pass without mishap, however, because we fell into the hands of a drunken coachmen who drew us hither and thither in a long roundabout way, so that we arrived too late to attend the marriage

---

theme of the play, a person who falls asleep like the Seven Sleepers did and in the final act wakes up to a changed world in a different time, realizing that the past was dreams within a dream, has appealed to many dreamers since Pedro Calderón de Barca's seventeenth century play *La vida es sueño* was wrought.

[58] A distance of 13,551 km or 7,317 nautical miles: passage time with a vessel speed of 10 knots was ideally 30 ½ days—with allowance for calms, rough weather, and the regular setbacks and backlashes making an average speed of 5 knots, and then well over 2 months of travel, which accordingly was shipmaster Söderström's and steersman Elfström's moderate calculation at the drawing desk. No wonder, that they had to set the whole crew at work with maintenance during periods when the sea was indeed *pacific*.

service proper. But the lavish wedding dinner we were allowed to enjoy from the start. The feast was pretty nice, or at least interesting, sort of. The bride and the bridegroom were particularly attention-grabbing to see. Mr Römcke, who now had entered the married state as husband, was pale as a corpse and looked heartbroken, frightened and on the brink of nervous breakdown—poor man!—beside his queen of hearts, who smiled with poise and looked motherly protective. The groom is actually only about thirty years old, while the bride is at least forty-five. Our father thinks that this marriage is a mistake, and that it cannot possibly be very happy.

Then again, I furthermore mentioned that we were also to a ball. It was in fact the same evening as the wedding took place. As it happened, a kind gentleman among the wedding guests politely invited us to the German Club's 30th anniversary ball.[59] Only the members and their families have access to the club, so it was formally a private ball. And it was a very gentlemanly thing. An Imperial Russian Navy manoeuvre is on out here now, and many of the officers were attending.[60] I was introduced to the Commander who asked for a dance, and [after polite brush-off, I] opened a conversation with the

---

[59] Several German-language clubs and groups have been operating in Melbourne since the 1850s, a time when many German migrants were attracted to Victoria by the gold rush.

[60] These naval officers were from the warship *Vestnik* «Вестник» (1880). Australia–Russia relations date back to 1807, when the warship *Neva* arrived at Melbourne during its circumnavigation of the globe and have since then mostly been tense. Czarist Russian men-of-war from time to time anchored at Port Phillip from the early days of the Australian colonies, and more frequent after the gold rush in the course of a long-term strategic plan for a Russian colony or base in the Pacific. For this reason, gun batteries were built in Victoria at both Point Nepean and Point Lonsdale to defend the entrance to Port Phillip Bay, as well as fortifications at Queenscliff, Portsea, and Mud Islands. On some occasions in the nineteenth century, rumours spread in Australian ports that a Russian invasion was impending. Fears of a Russian invasion where ignited when three Russian ships—the admiral's *Afrika*, the *Vestnik*, and the *Platon*—were sighted near Port Phillip in January 1882, and the war scare reached a new height in 1884–85 [Marmion, 2009].

gentleman, and chatted quite a lot with him.⁶¹ But he was pretty stupid, it seemed to me: he did not know Gefle and had no idea whatsoever as to its location! Confused, I think I made him believe that it was in Lapland.

Of course, Emy danced all the dances during the night!

Nonetheless I have given my God a promise, *yea*, a holy oath *never* to dance. It is my and my friend Lully's common promise to God. I have already made such a promise once before, and broken it. But with God's help I *cannot break it this time*, for I have seriously and solemnly uttered the words: God condemn me, if in this matter I am disobedient. I know that it is a sin to dance, even if it is not explicitly forbidden in God's word, which many people call attention to as a counter-argument. But my conscience tells me that it is contrary to the will of God and to the mind of the children of God. It's a worldly pleasure. The children of the world and the children of God have nothing in common, evil and good cannot be reconciled. Oh, that I always considered how important it is not to play with sin and act against one's conscience! God, come with your Holy Spirit, and give me a sincere mind. There is another delicate question I have, which is a point of conscience: is it right that I even accompany Dad and Emy to such places as balls and theatres, etc.?

---

[61] In 1886, the return of the large corvette-type war steamer *Vestnik* to Melbourne on 27 March 1886 (en route St. Petersburg–Cape Town–Melbourne–Vladivostok) was accordingly observed at length by the journal *Argus* on Monday 29 March. 'The Vestnik of the Imperial Russian navy, arrived in the bay on Saturday morning' heavily armed, it read, and with '170 crew members', headed by A. Baranov, Commander; Chief Navigating Officer A. Popov; and Captain Waldemar Lang; and carrying an impressive number of officers known from other naval operations, including the lieutenants A. Gorshkov, V. Zazarenny and V. Sarnavskiy. Of the crew, Andrei Alexandrovich Popov (Андрей Александрович Попов, 1821–1898) is particularly of note, a mariner known from the Crimean War. In 1886, the German origin of Captain Lang was a door-opener at the German Club ball—elsewhere these friendly visits were often surrounded by a charged atmosphere, and some places were too hot for Emperor Alexander III's men, who spoke German and French. Mia's *tête-à-tête* was with the Commander, not with the Captain or the Navigating officer, who were familiar with the major Baltic Sea ports.

But what shall I do, poor child? God help me to walk the right path! God, give me a right mind at the celebration of your Passover. Teach me to ponder, O Jesus, how much you have suffered for me, that your precious sacrifice may not be wasted for me, and perhaps through me also other souls. Oh, that would be awful!

But it is said in the word of God that 'he who does not gather with me scatters abroad'.[62] God in heaven, have mercy on your poor child and teach me 'to work your good works while it is day'. Teach me to bear in mind that the "Night is coming when no one can work." Jesus, teach me to love you more and more deeply. Show me more and more how you love me. Teach me to serve you, O my Master, oh, make me a humble and faithful disciple. Amen!'

11. Crossing the Pacific Ocean

'IN THE SEA OFF NEW ZEALAND, 6 MAY, 1886

On the go again! It was Easter Eve, *on the 24th of April, when we sailed from Melbourne*, and we have now been on the sea for almost two weeks (Figure 15). I want to summarize here, how our journey took shape during this time. We had excellent beautiful weather, when we left the port. We set out from Sandridge Pier early in the morning. Captains Swan and Lewis were down there to bid us farewell. They thought that they would see us again in America.

But because the wind was not fully favourable, we had to lie at anchor in the roadstead and did not put to sea until early the next morning, Easter morning. Therefore, we had the opportunity to go ashore once more. We spent our last evening in port at the theatre. The very Easter Eve itself! In this way we prepared ourselves to celebrate the holy resurrection of Jesus Christ. Praise be to God, that on Good Friday I still had the opportunity to go to church and hear something about Jesus, about my Saviour's cross, about how He is especially dead to me, because I am such a great sinner.

---

[62] Matthew 12:30; Luke 11:23.

Hence, it was with conflicting emotions that I went to the theatre. Nevertheless, as it is my intention to be fully sincere to myself in this my little diary, I would like to declare that I found myself quite amused when I got there. There was given a Japanese piece, called the 'Mikado', a remarkably well-liked piece that has been performed many times in Melbourne with much acclamation and applause. Brilliant music, magnificent scenery and costumes! The contents were actually nothing, but of quite a lot of interest: the event takes place in Japan, and it illuminates several Japanese traits and conditions.[63] Surely, I do think it is true that the theatre can often be good, useful and entertaining; it depends, of course, entirely upon the production and what is set up. The works of many good and famous writers are performed there, and where can these be better apprehended than in the theatre, where they are illustrated and stage-set! You can also hear fine music there, etc.

But the matter of decisive importance is distinguishing between the good and all the horribly bad things that are also performed on the stages, such as these disgusting French comedies, etc., which I get really sick by just thinking about them.

—Now enough about this and something more about our trip! It has been a cold and awful time these two weeks, especially a few days last week, when we had real hailstorms. We've been SOUTH OF NEW ZEALAND, so it has been very cold on board. And because the *Atlantic* is ballasted this time, we have felt the "rolls" quite a lot, and for a couple of days we have also become familiar with the common seasickness, although quite mildly. No need to worry for Mary, she is getting seagoing experience, and acquires seamanship. And, yes, we now have Mary Grundén with us on this trip.

We like her quite a lot, get along well, and she works comfortably with the crew on board. She is very quiet, though, far too quiet, I fear; but I

---

[63] By the end of 1885, it was estimated that some 150 companies were producing this comic opera in the world. When the Princess' Theatre in Melbourne reopened on 18 December 1886, this play was on the programme the opening day. *The Mikado* by Gilbert & Sullivan has been given in Australia almost continuously since 1886. Concerning the interpretation, this is a satire of British society in foreign disguise. Victorian audiences likely did not fail to catch this point.

hope she will feel better after some time, when we can talk to her more, and understand her better. This week we have been more conscientious in our English studies, and I want to keep up the good work. I hope that when we have been in the company of Mary and other English-speakers daily for about half a year, we will be fully fluent in English.

A few more words about Mary! About her appearance! She is neither ugly nor beautiful; still, she has pretty beautiful big dark eyes. She is a very sensible and capable girl. Her father praised her a lot. For three years now, since her mother died, she has been in charge of the household at home. She is the eldest of the siblings, nineteen years old. She has two younger sisters, Julia and Norah; these are very beautiful; of seventeen and fifteen years respectively. Then they have a younger brother of twelve. Their mother was Irish, consequently a Catholic, and all the children are Catholics.

We have now begun to long for the trade winds, and hope to have as good a time on the crossing, as we had on the previous passage. We have finally been granted with a little milder air.

7 May 1886

It is today exactly half a year ago that we left home. A thousand thoughts dash home today. Most of them with grandma as the addressee.

Today I feel like taking down a bit more about the acquaintances we made in Melbourne, so that I can then better remember them later. There are especially two Swedish girls, Bertha and Amelie Wohlfahrt, that I wish to recall, because they were so exceedingly kind and sweet. They are exactly of the same age as Emy and me.[64] They have been living in Melbourne only a couple of years, and they miss their home in old dear Sweden a lot. Oh, yes, I think everyone who comes out here must do so! I am also thinking about the considerate Mrs Zickerman;

---

[64] Amelie was born in 1866, Bertha in 1868. They were the children of Roberto Guido Wohlfahrt (1826–1899) and Malvina Carlsson (1830–1911) who had moved from Gothenburg to settle down in Melbourne, where they lived for the rest of their lives. Bertha was to marry Mr James Dickson-Waern (1864–1930), the Consul for Sweden, who actively supported trade exchange between Sweden and Australia. This family with connections worldwide became rooted in the world Down Under too.

it will be very nice and enjoyable to renew the acquaintance both with her and with the Wohlfahrt girls the next time we get here.

Mrs Zickerman, no matter how kind and nice she appears, there is a big flaw with her, it seems to me, since she is a freethinker who does not believe in God. She told us quite openly that "it would be real madness to believe in those children's stories in the Bible." That is absolutely awful! I imagined that she would feel unhappy, alone in a foreign land without God, but apparently doesn't do so.

Sunday, 9 May 1886

A rainy, heavy, and gloomy day! Several of the previous days have also been like this. We have had headwinds, and at times no wind whatsoever and got into a complete calm and become stagnant, unable to move forward, so that we are not getting anywhere.

Why does life seem so gloomy some days, that it really feels heavy to live? At such a young age! Yes, I am now closer to nineteen, and have many times experienced this. And the reasons? Sometimes it is only external, superficial, and insignificant disagreements that can completely weigh down my poor little heart. And that is the case even now. Alas, how dissatisfied and grumpy I am!'

12. '—It sometimes seems to me that I would like to rebel against myself, against God and the whole world!—'

Ugh! But why write this down, uh? Because it's true, and that I really am like this! If in these my notes, confessed to my confidante booklet, I only write down the good and happy thoughts that may possibly occur to me, (or rather, what God gives me through His Spirit, such as the eager desires I often feel to belong to Him!) it would be completely untrue to myself, and that I then received no true picture of the every so often fluttering, simmering, self-willed Mia S–m. I figure that it may be useful for me some time in the future to look back on moments like this, but then I also would like to be reminded here, of what perked up my sunken mood at this time, and that which revivified me. Yes, my futile hope would be to find here the best and safest remedies against heartburn and sentimentality, and the cures for sadness and troubles of all kinds!

Fortunately, I was not so deeply discouraged and downhearted this time that I could not summon up where I had sought and found help and comfort many times before. But I was *reluctant* now to open my Bible! Oh, so many times it has actually occurred lately that I find myself terribly unwilling to seek God in His word, that I thereby have postponed it time after time, and accordingly many days can pass again and again without me having prayed or read God's word, and as a result become restless and depressed.

However, today I read about "the good Shepherd". I read and re-read with increasing astonishment. Oh, yes, the heart cannot open enough to the riches of love that these words contain. To belong to Jesus' *flock*, how blessed! Jesus, my shepherd, and me his little helpless lamb! A stray, scarred lamb run wild, which He sought out, reclaimed, and healed!

I now remember a little verse, which I like a lot, and which fit exactly here:

> I am guilty, poor and weak, I know,
> But to your arms I flee.
> Be my power, righteousness
> My everything, Dear Jesus!

Yea, O Jesus, take me in your arms and lead and guide me through the *desert* of earthly life to the Father's home up there!'

They are now on passage through the South Seas at 25°4′ S 130°6′ W.

13. Pitcairn Island and the *HMS Bounty* Mutineers

'On 1 June 1886

Today we are passing the small island called Pitcairn in the Southern Pacific. How enchantingly beautiful it is, with its lush forested mountains, hills and valleys![65] The history of this island is extremely interesting. A small Christian colony has sprung up here owing to

---

[65] Surrounding albatrosses, shearwaters, petrels, and frigate birds in these waters add to the beauty of the Pitcairn Islands.

a number of amazing events. I would like to bring up something of that here.

The island was discovered in 1776 by a British warship, and was named after an English sea cadet who first got it in sight. It was then completely uninhabited. And after this time a couple of decades passed before the next time a human foot tread on its shore. Through the course of events, this island then became a refuge for a crowd of criminals, who here sought solitude to avoid punishment for their crimes. And this island, by its solitary, remote location, was particularly suitable for them.

They were namely sailors, and during a mutiny on board against their captain, they had sent their master and the few who stood by his side into an open boat to kill them all in this way. They nevertheless left them some water, and some bread and meat as well as a navigational compass, so that they could sustain their lives for a while, although all prospects of rescue for them seemed impossible as they were alone in the middle of the ocean, in the vastness of open sea.

And it did not take long before several of them died of disease and distress of various kinds. In this drifting predicament, they had in blank despair made land and disembarked at an unknown islet they had sighted, in the dawning hope of finding help there, but instead encountered wild cannibals, who overpowered them, and a couple were killed in the violent confrontation. It was with great difficulty the others managed to escape in their small boat.

In their desperate need, they began to pray to God for help and salvation. Perhaps they had never before in the days of fortune sighed to the Lord! But He is gracious, and He was so also to these unhappy people. After several weeks of wandering on the sea, they finally reached their place of refuge, an island called Timor, a Dutch colony.[66] There they told of their sufferings, and the horrible crime that had been committed against them.

The Dutch governor offered them help, and several of them then went home to their homeland, England. The British government

---

[66] This means that the deposed captain, and the surviving loyal crew of the full-rigged *HMS Bounty*, then had drifted some 3,500 nautical miles (6,500 km) east-north-east in an almost parabola-shaped course with the launch.

immediately sent an armed warship to the South Seas to search for the criminals, and several of them were arrested in Tahiti and punished after being summoned to uncover where their accomplices were. But they asserted that they only knew that after committing the crime, they had sailed to Tahiti, where they had lived a wild life in a debauchery for a while, haunted by the fear that their crime would be discovered and disclosed.

In actual fact, many of them had left Tahiti after a while, taking with them several islanders from there, both men and women, to seek a safer refuge elsewhere, and after this time nothing had been heard from the party.

But we have now learned about their secret whereabouts, and know where they had gone, namely to the small island of Pitcairn. But this intelligence oozed forth and was vented among mariners only after a quarter of a century had passed, and in the United Kingdom the whole event had long since been forgotten. But as it happened a British sailing ship on a voyage in the South Seas one day found itself right in front of a small island, which was not recorded on the sea chart they had. To their amazement, they found at closer scrutiny that it was inhabited, and they saw on the shore gorgeous, well-built houses, such as had not hitherto been seen on any of the barbaric South Sea islands.

Then a dinghy was launched from the shore, and approached the ship. It was sculled by two young men, and the astonishment aboard the ship became even greater when these greeted them in the English tongue. The two oarsmen were invited to a meal on board, and immediately at the beginning of this searching and tense repast at the dining table, they showed that they were Christians too, because they clasped their hands and offered a table prayer. Then the young men told who they were, and one of them was actually the son of the masters' mate aboard the ship *Bounty*, the instigator and leader of the mutiny proper. The other was the son of one of the sailors, of whom now only one still lived, their revered and beloved seaman John Adams; and their mothers were Tahitian women.

The officers on board the Royal Navy ship now went ashore to retrieve further intelligence and inspect the living conditions there. They found everything in the small community to be arranged in the most peaceful and pleasant way. They found that on the island there was a little Christian chapel, where the small congregation used to gather to hear the word of God—the aforementioned Adams was their priest and teacher.

But how could such a change have taken place with the wild criminal crew and their children? Yes, it had been caused by a copy of the Scripture, which the old Adams found among some articles without value that were taken from the *Bounty*. The very ship they had set on fire when disembarking, so that all communication with the outside world would be impossible and all traces swept away.

During the first part of their stay there, there was a terrible split and chaos in the settlement. A violent conflict broke out between the Tahitians and the Europeans, who later oppressed the former and treated them as slaves. One assassination followed another, and in the end all the male Tahitians were killed. In other ways, too, the culprits continued their criminal and sinful life. For instance, they started to extract intoxicants from a certain kind of plant, and thereby they naturally plunged themselves into even deeper misery.

But when distress and sin reached their climax, God sent them help. He aroused in one of their hearts the desire and longing for liberation and for a better life. It was then that Adams began to read in his Bible. God revealed himself to him through his Holy Spirit, and he found peace and forgiveness in the name of Jesus. He then considered it his duty to convey to the others this glorious message, and he began by teaching the children and women, and they embraced his words with much yearning and satisfaction. He taught them to read, and eventually a completely different life developed on the island, so that after a few years, when the English ship passed, the island was in this happy state. But that was a crucial and critical moment for Mr Adams.

He was now the one survivor of all the criminals. He confessed the whole affair and said: "Here I am now ready to receive my punishment." But when the report of all this reached England, he was pardoned, and several ships were sent out with European necessities and essential goods to the small colony.

The island is really very fertile and has an excellent climate, but due to its small size it was already too overcrowded, so that distress began to arise, and for that reason the British government sent aid with means for several of them to move to another island, Norfolk, to seek their better livelihood there.

Even today, Pitcairn residents continue to live a happy Christian life. It is amazing how the Lord God, through so much evil, has been able to bring about so much good, and this through a single Bible.

The Lord is mighty, and His word is a rock for all times.

Oh, how we wished when we passed this island that we should see some of its inhabitants up close. It would have been so terribly interesting. Thus, we were very delighted when we saw that a boat was embarked at the island and headed for the *Atlantic*. For that reason, we *hove to*,[67] and waited for it. But up close we saw that the vessel was so full of big, strapping fellows, about ten or eleven, that dad considered he did not want to risk receiving them on board. We had heard about so many incidents of how inhabitants of these islands have attacked and plundered passing ships. You can never be careful enough, and we could not be completely sure about the designs of these strangers. Dad furthermore said to us that even if their intention was good, it was still wrong and inappropriate to start out with so many against a ship. For that reason, we merely exchanged greetings with them and sailed on.

It was odd to see that they were all fully equipped in European outfits, two of them appeared to be dressed as real gentlemen. I suppose they felt disappointed when we sailed past them without approaching them any further.

Ascension Day [3 June 1886]

I remember that today a year ago, my dear friend Hanna and I during a long walk in the woods the early spring morning, spoke to each other about our joy over the glory in the awakening of all nature this season,

---

[67] Heave to: By this nautical term it should be understood here that the ship deliberately slowed down and came to a stop by turning across the wind leaving a headsail backed while keeping well off the rugged coastline and the reefs, yet hardly by taking in much sail in this case or dropping anchor in the shallow rocky waters surrounding the 'Bounty Bay' in waiting for the longboat—a kind of vessel still in use there, by the way! Pitcairn Island is only accessed from the sea at this bay [Irving, R.A. and Dawson, T.P., 2012].

and especially over the great importance of the day: that such day like this and Easter day must always appear differently than other ordinary days: brighter, clearer, more peaceful. At least it must seem like that to a Christ-loving mind. Alas, that I really had a true Christ-loving mind! Sure, I know that I love my Lord Jesus and that he is indispensable to me, but I often feel horribly cold and indifferent to these things!

But I want to flee to you Oh Jesus to find warmth and life. I want to find my all in you, my peace and my bliss. Because of your promises, which cannot fail, I will be sure of being received, and in this assurance, I feel happy and glad. I will go to you with my sins, with my coldness and unbelief. Through a little book, which I have recently read, I have been reminded of how important it is to truly leave the sins with Jesus. How many times has it not been so with me, that have I asked the Lord Jesus for forgiveness of my sins, and for the moment felt comforted and cheery, but then, perhaps within an hour, I have again in unbelief gone and sighed and been anxious over my sins. Lord, teach me to believe that you are a true and perfect Saviour, and that I too, through your inexpressible grace, may belong to you. I want to remind myself here of Paul's words: "Sin shall not have dominion over you, for you are not under law but under grace." Rom. 6:14.'

14. The Stars are Out: Under the Southern Cross

'Then again, I want to take down something more about this our Commemoration Day, and moreover to tell in greater detail about how our journey has turned out.

This has been a delightfully pleasant day, as a day on the sea in the tropics must be. Yes, we have now finally reached the much-anticipated trade wind belt, and are delighted with the glorious, warm, yet, as a result of the steady breeze, agreeably fresh air, not to mention the wonderful mornings and evenings! Of course, we have now started to rise earlier. Every morning we get up at about six or seven, before usually at eight or nine! And we go to bed very late, at eleven or twelve. At night we sit on quarterdeck or poop up in the stern and spoon in the moonlight, or admire the sparkling and twinkling starlight. We have begun to find again an old acquaintance from the north now, *viz.*, the CHARLES'S WAIN [the Plough in the Great Bear], but I do not know which constellation I admire most, this or the so-called SOUTHERN CROSS, which is so extraordinarily beautiful. God has set his bow in the sky as a sign that He is gracious; and the Cross, the sign of peace, the atonement, He has also set up in the south sky.'

Crux, the most familiar constellation in the southern hemisphere which includes the twain Pointers that indicate the direction of the south celestial pole—is in away the exact opposite of the Plough, the Pointers of which show the line to the Pole Star, the Stella Polaris.

The fact that both constellations have become visible to the crew above the horizon, shows that the ship has now moved a long distance northward from Australia and New Zealand, the national flags of which today carry the brightest stars of the Southern Cross as a symbol, as do those of Papua New Guinea and Samoa, territories passed by the Atlantic on the roundtrip.

Northern constellations visible from most locations north of the equator throughout the year include the Cassiopeia, Cepheus, Draco, Ursa Major and the Ursa Minor. Southern circumpolar constellations visible from most locations in the southern hemisphere are the Carina, the Centaurus, and the Crux.

The scenery of the sky above is changing for the travellers on their northbound journey, now turning to astronomical navigation, and who are spellbound by the sight when the universe opens at night.

During a starry night in the southern hemisphere, you can see straight into the centre of the Milky Way, and that is why you can see so many more stars there than in the northern hemisphere. After nightfall, the universe is put on view and the scope broadens indefinitely. If the firmament had appeared only once, there had been legends worldwide of this singular marvel, on how the way of the heavens had been visible in a blessed moment in the past, and how the peoples on earth had been granted to behold the tenement of the infinite and the celestial origin of our species. Some autochthonous peoples perceived this. No wonder then, that the lonely navigator at the steering wheel on night watch, guided by the stars since times immemorial, is awed by the scenery and grateful for the order of things.

'As a contrast, our month of May was for the most part very unpleasant, *huh* so cold and terrible it was, and as a result of the constant HEADWIND we drifted farther and farther south—it would have been interesting if we had reached the South Pole![68] It was with quite a lot of jealousy and longing we then thought of them at home, and of the glorious May days we had previously experienced, but now I feel deeply satisfied and happy. It is also only when it is foul and adverse weather and when we have the wind dead against us, that we long for entering a harbour; on the sea in weather like this, though, it is a thousand times nicer than it can be in any anchorage!

Mary agrees, and she is doing really well here on board. I am assiduous in my English studies. This week, however, I have not had time for that, because we have had piles of laundry to undertake, and two days we were busy with ironing, another day with darning and mending of socks, another day with hanging out clothes and brushing them, etc. We are never without employment and tasks on board, and therefore time also passes quickly.

We now also have got the opportunity to take a bath every day, since the carpenter has built an excellent comfortable bathing cubicle.'

During the months spent on crossing the Pacific there were calmer DOLDRUM periods mid-ocean, and hence very useful spare time, on which the master set all the crew to work with various neglected duties, including domestic business that could not be attended to at other times. The skilful and always busy carpenter PER JOHAN BLOMGREN, a married quadragenarian, manufactured a little hut so that the women aboard could have a wash, too, out on the open sea in the protracted time of ocean passages.

---

[68] They reached the 46th parallel South, on line with the French Southern and Antarctic Lands but saw no ice.

15. 'Love without Expectation'

'15 June 1886 [Tuesday]

We have now celebrated our Pentecost. It has passed very calmly and quietly. Emy and I have reminded each other of the Pentecost two years ago, when we were confirmed.

Yesterday [14 June], the second day of Pentecost [*Whit Monday*], we had a dreadful rainstorm all day. To cheer me up a little, I seized the opportunity to read through the dear letters we received in Melbourne. In them I was instead informed about the engagement of "*a certain person.*" Oh, that I could receive this news so wondrously coolly! There was certainly no need for a "sofa corner or Eau-de-Cologne bottle," which Lully so thoughtfully recommended when she informed me of *yet another* person's engagement. But as I said, I took it easy. My hopes in that directions have long since been dashed and shattered. So now my youthful dream is over, and everything is just a memory, which sometimes seems bitter to me, at other times ridiculous. A fire, if it were ever so fierce, must sooner or later go out and die down if it is not maintained, and so also mine! Poor me, whose lot is to love without expectation! Yet, in blank despair, I nevertheless cherish hope! And henceforth – never more a word about this thing!'

16. Mid-Ocean in the Intertropical Convergence Zone

—'first birthday with daddy'

'We crossed the *Line* on THE 9ᵗʰ OF JUNE.

Tomorrow [16 June] is my birthday.[69] I wonder so much, how it will be celebrated. Everyone, both Dad, Emy, and the steward, look so mysterious, and I have surprised them a couple of times, as they were whispering about something.

---

[69] She was born on the 16 June in 1867 and celebrated her 19th birthday now in the doldrum region mid-ocean above the underwater Cooper ridge at the seventh parallel—the first anniversary in company with her always absent father. To catch up with a sea captain, you had to join his crew.

16 June 1886

A very memorable day! I have never in my life been so celebrated as today. This is also the first birthday, which I celebrated with Dad. In the morning I woke up to music, and when I opened my eyes, I was surprised to see a big strange old man sitting in an armchair next to my bed. At first, I thought it was Emy or Mary, but the next minute they came in through the door also fancy-dressed as "old men"; Mary rigged-out to sailor in white trousers and a blue woollen shirt, slouch hat, and pipe in the corner of her mouth, and Emy dressed up as a gentleman "dandy" wearing tailcoat, white scarf, and white gloves with a walking stick, pince-nez, cigarette, blackened moustaches and eyebrows. Dad was sitting in another armchair, dressed as Neptune. He sat there motionless, in the same position as the man next to my bed, but the question now was: "Who *is* this?". I was almost terrified, half-awake as I was, and had to ask, "How many are you?" ("Four people here, he's possibly the mate, but that would be rude!" I thought, pulling the blanket over me better.) But then Dad and Emy came to me with a huge parcel each, and I had a terrible trouble opening them, and getting all the dozens of papers that were wrapped up. Eventually I found in Dad's package a small matchbox, and I was about to throw it away, but luckily I anyway opened it and to my great surprise found three English pounds inside it and a piece of paper with the writing: "Congratulations from Dad on the 19th birthday! Basic funding for a bracelet!"

Emy's parcel contained a beautiful white string of beads.

Then I was treated to coffee in bed, with plenty of "dipping" [buns], which Östling [the steward] provided. After I had dressed, I was escorted out on deck by the two young gentlemen. Flag was hoisted in my honour. Thereafter the morning passed calmly and quietly, and we had our regular work to do as usual. "My old man" in his chair, they had moved out to me on deck. And I had by then realized that this male figure was *stuffed with oakum*,[70] but indeed successfully true to

---

[70] Packing tow used for caulking the ship: a loose-fibre waterproof sealant, tarred and stuffed into the seam of the planking with hammer and iron. This sealing had often to be mended. Traditionally of hemp—the same stuff as the strong multiple-twinned cordage in the rigging.

real life, and an illusory effigy that took me in (and worried me a bit). They had now also given him an empty bottle to hold on his arm.

But after we had dinner, the most comical thing of all day happened. During dinner, I began to suspect: "now there is definitely something going on," and I managed to persuade Mary to admit that the crew had all afternoon free to have fun. So, without further ado we went out and took a pew in our seat amidships, where the "old man" was still sedentary in repose. But after a few seconds, Emy exclaimed: "Hey, look! This is a living human being!" She claimed that the shape of the head was not exactly the same as before. We began to touch him to find out; finally, Emy pinched him in the nose, and then he got up and bowed very neatly to me, and we recognized the [first] mate [*Elfström*]! It was him!

Only a short while later, an entire caravan[71] came marching in procession with a big barrel organ on wheels in the lead, which was certainly very heavy, since several had to pull it. Then we wondered, where that musical instrument came from. However, it turned out that a man was sitting inside it, playing the accordion. Little Pussycat was dressed up as a monkey, and *Harriet*[72] as a bear, and she was led by an old humpbacked man at the end of the parade. Everyone wore the most amusing costumes.

The conductor of the pageant turned to me and expressed "everyone's congratulations to today's heroine," and they shouted a triple "hurray!" Then they sang a few songs, and next they began to dance, and the dance on board continued until late in the evening. We watched and laughed a lot at all their funny shapes and guises in the priceless suits. At coffee time, a deeply veiled lady with the coffee tray and a large birthday cake with "19 years" comes out. It was almost impossible to recognize who it was. I wondered if it was possibly the steward,

---

[71] Originally 'carnival', later changed by the diarist.
[72] Harriet (Figure 11) was the name of the *Atlantic*'s most popular bitch, a long-lived watchdog who kept a check on rats and birds, and in old age attracted all lice. She was the best swimmer and lifeguard on the tour. The ship dog and the ship cat were cherished companions of many sailors. In earlier times, the words 'dog' and 'cat' were *taboo* on board, but for later generations both were regarded as auspicious. They were considered to be able to predict impending danger, and if they ran ashore just before departure, it would be wise follow their example and abscond.

but I had seen him a little while ago, and he always has moustaches. Of course, her voice was muffled and squeaky as much as possible, but once, when "she" laughed, I recognized the steward's voice. And he had really shaved off his moustaches the minutes before to look like a real lady!

Everyone on board was well entertained and catered for, they had all a good time and were all very happy about their day (Figure 11).

Mary says she always wants to remember it, and so do I.

In the evening, when we sat up on quarterdeck and enjoyed the moonlight and talked about all the fun that had happened during the day, all the sailors came up there and thanked Dad and us for all the nice things they had, and once again they congratulated me, and *hip, hip, hurray!* was ejaculated again.

In this way the day ended and is now but a memory, like everything else that is fun or sad that have passed and is over.

Anyway, the account of this day is a clear proof that you can also celebrate a really nice birthday on the sea in the middle of the open ocean.

Sunday the 27th of June

Today a hot battle has been fought within me, and with God's help I have emerged victorious.

Oh, it has now been about a special way of believing without seeing with one's own eyes and feeling, beyond the testimony of the senses! Is it really possible that God can now look down on me in mercy and compassion? Do I really now have the forgiveness of my sins for Jesus' sake? Yea, it's possible; I believe it, and you, the Holy Spirit of God, have convinced me of this.

I do not want to doubt my dear Master's words and promises. They are so many and precious. They apply especially to me, because I am such a great sinner. Jesus has countless times invited me to come to Him and to remain in Him. I myself have not been able to come, or do the slightest thing to be preserved with Him, I know all that too well. Every day I am convinced of my own inability, but the Lord Jesus has drawn me to Himself, and now He also wants to do everything to keep me with Him.

The reason why it has now been so difficult for me to believe this, is that once again I have had a violent outbreak of sin. Last night I fell out with Emy when we started arguing, which, unfortunately, occurs more often than not, although after a while everything tends to get better again between us (Figures 12-13). But this time I let fly at her and railed with such frenzy that there were no bonds to my anger, nothing could stop it, the fiery fury boiled over, and in the eruption overflowed all banks. The next moment, I was sorry, repented, and I was seized by a terrible despair after the brawl.

This thought then struck to me: "Is *this* the way I profess my Master?" Many, many times before on similar occasions, I have for a long time, yes, several hours or days surrendered to this despair, without even daring to think of receiving forgiveness from my God. But this time I felt a clear exhortation: "Go, cast yourself down before your God, and immediately pray for forgiveness!" And so I did, while wrath still boiled within me. Alas, it was certainly not a coherent prayer! Only sighs and an incessant utterance of: "God, my God, save me, have mercy on me for the sake of Jesus! Is it true that Jesus saves me? Is it possible that you want to forgive me? Oh, God, teach me to believe!" And I tried to convince myself that it was possible. God's Holy Spirit reminded me of the precious word, "Jesus receives sinners,"[73] and I said this over and over again to myself, and that, "Jesus saves me, Jesus saves me now," until I was finally filled with peace, joy, and thanksgiving for this wonderful truth.'

17. Port Townsend, USA

—whales and sea devils in sight

'THE 14TH OF JULY, 1886 [WEDNESDAY]

Port Townsend, Washington Territory, United States of America! This is currently our address! We arrived here on Sunday, i.e., on the 11th of July. We just did not have a swift trip, we arrived at least two weeks later

---

[73] From F. Engelke (ed.), *Lofsånger och andeliga visor* (Stockholm 1875), a partly Moravian collection of psalmody.

than we should. But we attribute this delay to the constantly adverse winds at New Zealand.'

They arrived at the northernmost part of the American West Coast on this trip from Australia after seventy-eight days—that is, almost exactly on schedule according to the captain's original and very realistic computation of time for a diagonal crossing of the Pacific, i.e., lasting about two and a half months. They set off from Sandridge Pier in Melbourne on 24 April 1886. The ballasted freighter Atlantic could make a good ten knots or more in favourable winds and currents daytime, but the average speed during this long voyage was five knots around the clock, and there were indeed much adverse winds to begin with.

'However, now when everything has gone so well, we should be thankful to God. And I'm really happy and grateful now. In particular, I am delighted with all the good news from home. From our little grandmother, who is daily and at every moment in my thoughts, we have received long, lovely letters. It is as usual with her health; alas, that she did get better from that nasty gout!—otherwise and in other respects everything is well and satisfactory, and best of all is that she moves to Aunt's farm this autumn.[74] It's so comforting to know this.

I thank you, dear Lord God, for fulfilling this our sincere desire.

By the way, everything is fine with all people at home.

I now have a lot to note down on what has happened in the last few days, and it will be done with difficulty as it is hard to remember everything. I now use the first free half-hour to be still and gather my thoughts.

---

[74] Their mother's foster mother, the orphanage founder Elisabeth Charlotta Leufvenius, then frail and about 76 years old, was the widow of the early deceased sea captain Olof Winroth, and in 1886 she moved to the house at Arbetarhusgatan in Gefle owned by their paternal aunt Catharina Charlotta Söderström, the widow of sea captain Pehr Zellinger who was lost at St. Helena.

Oh, would that nothing may draw me away from this One and only, which is worth seriously attaching to the heart and mind! Jesus keep me with you, this is my daily prayer!

What an exciting and awesome area this is! No words can describe how wonderful this is. And yet I cannot refrain from mentioning the soaring, forested mountains, some towering peaks of which are so high that they are covered with eternal snow.

It has been especially joyful and surprising for Mary to see the snow, she has never before seen any. For our part, all this has reminded us of our dear homeland, and even more so the fresh and fragrant coniferous forests. I come to think of Grandma, when I see them, the forest is her delight, oh, what a forest walk would make her happy, if it were possible! Now she still has Auntie's little garden to sit in, and admire the flowers, her darlings. She calls them "her children", she writes, and "next to our letters, which are her greatest joy in life, it is her flowers"! Poor, poor little darling!

It was also very fascinating this time to reach the port. We saw the first glimpse of land already on Saturday morning [10 July], although we were then quite a long distance away, but the land here is so immensely high. For several days off the coast, the whole sea was full of WHALES.[75] *It was the first time we saw any*, so it was very interesting. We then also saw a strange, horribly ugly fish called the SEADEVIL.'[76]

---

[75] Eye-catching, to be sure, since this might have been species of *Balaenoptera musculus*, blue whale, still common at that time, probably the biggest animal that ever existed. Moreover, there were the Grey and Humpback whales, and the small Minke, all often seen here off the coast, along with Orcas, killer whales, occasionally in great number. Still today, in 2022, there are Puget Sound whale-watching tours departing from downtown Port Townsend. Later on, they would see Spermaceti whales (*Physeter macrocephalus*) off the east coast of South Island in New Zealand, the most profitable game and favourite target of whalers.

[76] In early nineteenth-century Swedish, *sjödjefvulen* ('the sea devil') was sometimes connected with the fresh-water *Näcken*, 'the Neck', the evil spirit of the water, which had plenty of counterparts in oceanic monsters of various kind and size, all reminiscent of Old Nick one way or another. The author of the diary may refer to a large animal, such as the devil ray (genus *Mobula*) or a manta, which they might have

18. Among Rothschilds and American Indians

'Sunday morning [11 July] we got up early, already at 4, and all day we did not get tired of admiring the wonderful sceneries. Port Townsend is located 70 English miles up the strait. The strait forms the border between the United States and the British colonies.[77]

We stay here only a few days to unload the ballast, and go to load to another small place, Port Gamble, a couple of hours further up the strait. I think we go there already today in the afternoon.[78]

We have met a family of brokers named ROTHSCHILD. Their daughter, *Regina*, is of the same age as us. We have been in her company daily, and we like her a lot. She is brisk and sprightly, a truly American woman, and she is really cute, we especially admire her wonderfully beautiful gold-coloured, curly hair. We have had a great time in their large garden, it happens to be strawberry time now! We've already had a forest walk, and we thoroughly enjoyed it. My little friend, *Hedde*, in a letter expressed her regret that I had no woods to walk in, as she has at Mamre! But now I do anticipate spending a lot of time in the forests, indeed.'

---

observed from the rail; or the much smaller, truly infernal genus *Melanocetus*, black sea devil or humpback angler fish, of the family Melanocetidae, which indeed has a repugnant countenance. The point of issue then, is how they got such deep-sea creature on hook or in their net.

[77] The Columbia District in British North America, bordering to the for long disputed Oregon territory. Today, the recognized border between the USA and Canada runs through the Juan de Fuca strait or channel that extends from the Pacific Ocean to the Puget Sound.

[78] Port Gamble, with access to ocean commerce, was founded as a Puget Mill Company town in 1853 following the goldrush and was in the 1880s a famous port for exports of timber and wood products from this rich fir forest region—and that was exactly what the captain of the *Atlantic* and the agent C. Gunnersen were after: to replace the ballast of marketable corn in the deep cargo hold and be sold on return to Australia. This being so, the Melbourne *Argus* issued a report on Friday 19 November 1886 about the arrival of the *Atlantic* from Port Townsend to Hobson's Bay, carrying '10,150 pcs [pieces of] (800,000 ft.) *lumber*, 1,725 bdls [bundles of] *pickets*, 135,000 *laths*', all cleared. That was gold in the building boom of the barren Australian wastelands. Gunnersen was one of timber-importer Römcke's business connections.

The businessman and shipbroker DAVID C. H. ROTHSCHILD, *nicknamed 'the Baron' had died at 61 shortly before—on 24 April 1886—leaving his wife Dorette, née Hartung, and four children, including* REGINA ROTHSCHILD, *who was born in 1867 as was Mia. This daughter, married Jones, died in 1942 and was buried at Laurel Grove Cemetery in Port Townsend as her father and other family members were. When the crew of the* ATLANTIC *was visiting, her brother Louis had taken over the business together with their mother, who remained in the Rothschild House until 1918. The property was listed on the National Register of Historic Places and is open to the public today.*

'Something very interesting that we have seen here is INDIANS. We have even paid visits to several of their little huts. It was horribly untidy there, and they themselves looked like they had never washed themselves or combed their hair in their lives. But they were not as terribly ugly as I had imagined, quite on the contrary, and the children in particular were really beautiful with their big black eyes. I felt very sorry for those poor people. I asked Regina if the priests here did not talk to them about God, and she said that they did not want to listen to it, but that they were forced to send the children to the Christian schools.'[79]

---

[79] The number of different American Indians in this region had long ago decreased to small groups due to diseases, wars, and oppression verging to extinction, living in small huts and hovels more often than tepees, and supporting themselves from traditional hunting, fishing, and gathering. A few have survived and multiplied to some extent, such as the Port Gamble S'Klallam Tribe at its reservation, but the Christian European settlers of the Americas have nevertheless been guilty of genocides. Native Americans have been oppressed, moved, their settlements destroyed, hunting grounds and holy mountains exploited by intruders, but not all have been involved in wars, defeated, or vanished, however. There are today twenty-nine federally recognized indigenous tribes throughout Washington, including the more prosperous S'Klallams the sisters met with, and the *American Library Association* adds three more: the Duwamish, Wanapum, and Chinook, which have had a long history in this state. In Port Townsend, Port Gamble, Puget Sound, and Little Boston in Kitsap County, the crew of the *Atlantic* appear to have encountered some S'Klallam, headed by Chief Chetzemoka; along with Skokomish, and a few lonely Chimakum.

19. PORT GAMBLE—A Meeting Place for Travellers From Afar

'PORT GAMBLE ON THE 17TH OF JULY [1886]

The days we have been here, we have been living as real "*forest people*". And with that, we intend to continue as long as we are here. It puts life, desire and joy into a human being to stay in this forest. And, what a forest! It is richer and denser than the forests at home, and perhaps this is what makes it more beautiful, something so wild and magnificent. It gives us an idea of the American primeval forests. Nor can we here as at home—as a result of its density—wander across it between the trees.

I think that the nature here is more similar to Norway's than to Sweden's, but somewhat more southern in character. The *cedar* grows wild here, although not very high. It smells so nicely. But what grows tall here is a kind of palm-like ferns, they are as tall as the spruces, and meander around them. Nightingale song we hear here too. Oh, yes, everything is so above all description wonderful!

Our trip up here from Port Townsend was also very pleasant. A very small party was gathered on board. In the morning I had been ashore and shopped a little in the company of Regina, and both she and her brother [Louis] accompanied us here. So did also a couple of young ladies, which are living here, they were all with the steamboat, which was towing the *Atlantic*, but we invited them to come on board to us. People here are very nice and hospitable. The women assured us that we will no doubt have a very pleasant stay here, and of course will be introduced to all the "fashionable" families and be part of their pleasures, such as the so-called *pic-nics*, which they often enjoy themselves with during this summer season. Well, that will do, but whatever turns up, the forest walks will be the foremost pleasure. We are now invited to a big ball, which will be on Tuesday. Would I could find a way to be excused from that and avoid it!

Last night we had a nice sailing tour in the company of those wives and their families. A young Norwegian named Kildahl was also present, so that we could speak our own language. By the way, we now speak English fairly well; I can say to our praise, that we can understand and express everything. This afternoon we should be accompanied by the young wife of an American sea captain, Mrs Kilman. Their ship is next to the *Atlantic*.

She and her husband came for a visit here aboard today.[80] She said she was lonely and bored and found it so tedious in this place that she wanted to be with us. I do not know why, but we do not really like her, she seems so affected, and I do not think we will get along well. But it is probably not right to stick to the first impression.

Now we do not like Mary at all. She's getting so pretentious and disagreeable. She is also terribly sluggish, and a word she often uses, *stupid*, may well fit in with herself. Every so often we are really unhappy in her company. However, we will try to keep the tide up and maintain harmony.

And in any case, it's stupid of me to write down my unfriendly feelings like this.

A rebuff occurred to me this week. God, how desperate I was! But you helped me, dear Lord God, and all is well now. I really do not want to write down here what it was all about, I will probably remember it anyway, although I wish I could forget it. Yesterday I had a letter from Maria Werner: Doris is now engaged to her dear pastor Broberg, as I predicted several years ago. My poor little Maria, she is so inconsolable. Alas, if she could all the same get over her heartache and the wound self-heal, it must have cut so deep.[81]

TOMORROW, SUNDAY, we want to go to church, and this will certainly be really good and fine.

Another Sunday, though, I would rather like to go to the Indian Church. A French Catholic priest is a missionary among the Indian population here, and a small white-washed church has been built in their small community on the other side of the bay here. There are about 50 small huts on the site; that part of Port Gamble bears the

---

[80] Her husband was reportedly lost three years later.
[81] Maria Werner's sister, Doris Rosina Werner, had married pastor Karl Gustaf Broberg in Gefle. *A crime passionel.*

stately name of Boston.⁸² The Indians in this place are quite civilized, many of them work in the sawmill and earn good money. They also have nice and tidy homes.

This part of America, I think, is a gathering place for people of all races and colours. It is also full of Africans and Chinese. These also work in the sawmill, and the CHINESE, these "sons of heaven," as they call themselves, have, of course, here as in Australia, all sorts of occupations. A very good laundry is kept by the Chinese here, and we intend to send our clothes there, although it is not very pleasant to let Chinese gentlemen wash our intimate wear. They also go around selling a kind of berry called *blackberries*.⁸³ We always meet them in the woods, coming with their baskets. They always have the custom of following the roads walking one by one; if the road is never so wide, they never go abreast. This looks very ridiculous. Another peculiarity is that all Chinese are called by the name *JOHN!!*

Blackberries are very tasty, somewhat similar to wild raspberries⁸⁴ at home. One day we intend to amuse ourselves by following the steward to the forest on picking.

The one who perhaps most of all enjoys this forest life is HARRIET [the ship's dog, see picture of the crew]. She is overjoyed. This is different for her than in Melbourne, where she was never allowed to go ashore, but sometimes even had to be in leash, poor thing!

---

⁸² Little Boston at present-day Port Gamble Reservation, which includes the tribal communities in Kitsap County across the so-called Hood Canal (which is a natural waterway). Seat of the native SUQUAMISH (the people of 'the clear salt water'), which still in 2022 celebrates its heroic chief *Seattle*, who died a Catholic in 1866, lent his name to the seaport city, and is remembered for his speeches on the preservation of nature. The Söderström party, along with the Rothschilds and other settlers and inhabitants, is now living in New England-style wooden houses on the thickly wooded Olympic Peninsula, one of the last explored territories in North America and not yet fully mapped in 1886. The Olympia had become the major centre of forestry and felling in the region, to the detriment of the Skokomish and other aboriginal natives, indigenous to the state, who shot by shot were sent to 'the happy hunting-grounds'. There was indeed a Catholic Indian Mission, and the Shakers erected their first timbered Indian Shaker Church in 1885.

⁸³ *Rubus fruticosus* L. Wild blackberry, small sweet fruit native to the temperate regions of North America on the Pacific coast.

⁸⁴ Arctic bramble.

On Sunday the 18th of July, there was no church service, which was a mortification. Instead, we were out on a *picnic*, to which we were invited by a family named Thomson. They are happy and nice people. The wife is young, only twenty-two years old, very good looking, and coquette and flirtatious, too. They have two small children, a boy and a girl, who were also on the excursion. Then there was a sixteen-year-old very beautiful girl, Miss Guptill and her brother. Furthermore, Mr Kildahl, with whom we have previously become acquainted, and so us, "the whole family" from the *Atlantic*. Steersman Elfström was also present. We had quite a nice time. One little circumstance deserves to be noted, viz., that Mrs Thomson forgot the foremost food hamper, so that Mr Kildahl and helmsman Elfström had to sail back to fetch it, and we had to wait a very long time for our dinner together. We have already found that we like Americans much better than Englishmen. They are dashing and full of "go," and you always feel at home with them.

In the evening, Mr Kildahl and one of his elder brothers, who lives in a place nearby, came aboard. After drinking tea, we went for a forest walk in the moonlight.

See there how we spent our Sunday!'

20. Forest Walks

'Monday and Tuesday mornings were then spent writing letters. A letter from my friend Hanna has made me very happy.

In the afternoons we were of course in the woods. We are on the ship as little as possible, because we have a rather unpleasant anchorage here, close to the sawmill, so that it is terribly smoky and dusty all days.

Usually we take our work out with us, reading matter and refreshments as well, such as fruit, cakes and juice and water. We really feel very well, and cannot possibly wish for anything better. It is also quite romantic, too, if we consider that quite close to the place in the forest where we usually are, several Indian wigwams have been set up. We see the children running around there, like shy hares.

What else do I have to write about the rest of this week? Tuesday night we were at that ball, but there is nothing to say about it. We were introduced to quite a few people, to the head of the sawmill and his family, etc. Emy and Mary danced some dances, but several of the dances here are not at all the same as at home or in Australia and unknown to us, so they were sorry not to be able to dance them.

On Friday [23 July 1886] we were out picking berries. We sailed several miles further up the Sound and landed in a very beautiful place in the forest. There was a small farm nearby, and the edifice with its small garden looked so authentic Swedish. We walked on the lovely little path, and it led to a spring of fresh, crystal clear water.

The berry picking was a great success, we got several jugs of berries. Our skilful steward has now made jam of them in the cookhouse kettle, and it is "exquisitely delicious," our father reports.

Sunday the 25th of July

I'm terribly sad that it's going to be impossible for us to go to church for a whole month. That is so, seeing that no priest is permanently resident here, but travels at certain times to minister at different places. I was close to crying when I heard that. The heart becomes so cold and dead by never being warmed by His word in His own holy house. I hope, however, that one day we will be able to travel over to Port Townsend, where we are invited to Regina, and we will then be able to go to church.

Today we have been on a similar excursion as last Sunday, but in much larger company. Among other things, they amused themselves by swimming and fishing. But I did not have much fun. In the evening, the whole party came here on board. I was terribly tired and sleepy.

Monday the 26th of July, 1886

Emy's birthday![85] We had plenty of merriments all day. And Emy was celebrated and honoured, indeed! The whole *Atlantic* was decorated with leaves

---

[85] The younger sister Emmy Axelina Charlotta was born on 26 July 1868, and thus celebrated her eighteenth birthday.

and wreaths on deck, in masts and yards. A very large garland in the shape of a medallion with the number "18" in the centre was placed straight in front of the cabin doors. She was awakened with cheers at sunrise. Of course, she received the same kind of gifts as I got on my birthday and a lot of sweets as well. She was also given three fine perfume bottles, of which she gave Mary one and me the other one. Mary and Emy agreed upon not to open them until the morning of their wedding day. But I opened mine immediately, and invited them to share it, in honour of the day. In the evening we had quite a few people gathering to party on the ship, among them two young ladies and two young gentlemen, who were very nice and jolly. Furthermore, Captain Cutler with his wife and their little daughter, Ruth (Figure 14). These are a really very charming and amiable family!'

Fig. 14. *Captain Cutler and his daughter.* Courtesy of Colin O'Neill.

Captain ROSEWELL DWIGHT CUTLER was born on 29 March 1845, in Maine, and during his long life on the sea as sailor in the U.S. navy as well as the merchant fleet he was the master of the barks ARKWRIGHT, the CAMDEN, and most legendarily the barkentine KLICKITAT on which he took command in the beginning of the 1880s. Lumber laden, he regularly crossed the Pacific from Port Gamble to the Sandwich Islands.

On 2 September 1900, it was communicated that he had arrived with his ship from Port Townsend to HAWAII on his sixtieth roundtrip, this time with the mainmast broken after having been caught in a gale (THE SAN FRANCISCO CALL, page 28), which was also announced in other newspapers. On 17 April 1908, it was reported in THE HAWAIIAN GAZETTE (Honolulu) on page 3 that he had arrived from Port Gamble with the barkentine KLICKITAT to Honolulu on his eighty-second voyage to the port, fifty years since he first went to sea.

The Gazette depicted this old salt as tender and kind 'as a woman'.

He was married since 1879 to ALURA ELIZA BROWN, born in Honolulu, the daughter of the whaler captain Robert Brown. A son, Gilbert, died at his birth, on 20 February 1886; which means that the Captain's family in July was probably still going through grief from the loss of their baby.

The little daughter RUTH died of typhoid fever in 1899 at age 15. Captain Cutler died in 1916, in Seattle, Washington.

THE DIARY CONTINUES

'TUESDAY 27 JULY we were out for a sail in the company of Mrs Kilman. We went to the same place as last time, and we picked a lot of berries. The berry picking is quite tiring, I must say; we get used to crawling and climbing over fallen tree trunks, which are burnt and sooty. *Blackberries* grow here in profusion, above all where the forest is burned. And then, we have the terrible mosquitoes to fight with. But – there's no fear of that, we are young, hearty and brisk! And exercise does not hurt!

This time we had the honour of getting to know the family on the small farm, which I mentioned earlier. The "Lady" is a Native

American, the husband an American, the children are almost white and look quite nice. We were invited to their little *parlour*, which was very nice and decent. The family was extremely kind to us, and gave us bouquets of flowers and a basket full of vegetables.

When we got on board in the evening, Regina was with us to stay a few days in our company. She is really a little madcap, and there was a talking and laughing almost all night, we did not fall asleep until at three o'clock in the morning, running out on deck and into the rooms undressed.'

21. Little Boston and Day Cruises

'Wednesday morning, Mrs Kilman came to say Goodbye. We have been a lot in her company all the time, and learned to like her more and more the more we got to know her.

In the afternoon, we walked in the woods. We settled down at our habitual favourite place. This time I was in repose and had a very peaceful time alone with my God, while the others went into the woods for half an hour or so and picked berries.

In the evening we went cruising in the company of Captain Cutlers. We went over to *Boston*. There we got to know a little more about Native American manners and customs. The elders still want to keep to their ancient nomadic habits. Huh, we looked into a couple of their terrible sheds, these cabins could not be named otherwise, and within there were whole families packed together; the fireplace was in the middle of the floor, and a hole in the ceiling formed the chimney, and the whole hut was filled with smoke worse than in a buckling [smoked herring] furnace. In some other cottages around, however, it was nice and fine.

We entered their church, visiting both their old pagan cemetery and the new Catholic one. The custom used to be that all the belongings of an Indian should be buried with him, some things they had also hung outside the graves, such as axes and other weapons. The new cemetery was located on a mountain and an august beautiful path led up there. One of the Indians accompanied us, and he showed us the graves of his wife and his children. They have recently died and it was really very moving to hear him talk about it.

THURSDAY excursion and berry picking again. This time we picked the berries for Regina. I got my poor hands so torn. We are terribly tired after our trip. Our pleasures come too close to each other, so we do not have time to rest between each time.

ON FRIDAY, I accompanied Regina to a couple of families for visits. In the evening a large company of young people gathered here on board, and it was a tremendous atmosphere. Among other things, we girls found a way to deal with the moments when the gentlemen went out to another room so as not to bother us with their cigar smoke: then we also took cigarettes each and all smoked so much that the room was full of smoke. Our strangers left at midnight, but we did not have the pleasure of going to bed, but had to be up until 3 a.m., when Regina left. It is indeed inconvenient and stupid that the steamboat leaves at such an early hour, in the middle of the night. Daddy was also leaving then for Port Townsend to see Mr Rothschild, Regina's brother [Louis, who was taking over the shipping business after his father's recent demise].

SATURDAY seemed very short, due to the fact that we slept until noon, which we also needed after all the night vigil. At three o'clock, we went down to the steamboat and fetched Dad, who then came back. In the evening the Captain Cutler's were here.

SUNDAY THE 1ST OF AUGUST, 1886. Messrs. Kildahl were with us all day. We went and watched a prize competition in handling a ball in a regular [American] football match. The contestants were a number of gentlemen from PORT LUDLOW [Jefferson County] and natives from the SANDWICH ISLANDS [Hawai'i]. The Islanders won.

In the afternoon, several young people came here. One of our girlfriends here, Miss Eunice McPhee, we like especially. The Americans are generally extraordinarily "handsome", but they sometimes seem to me a bit too free and easy. At home it is usually the fashion that in company it is the gentlemen who "entertain" the ladies, here it is on the contrary the ladies who lead all conversation, and do everything so that the poor gentlemen do not feel too timid or diffident. Captain Lewis, our old acquaintance from Melbourne, has arrived here.

From Captain Swahn, who at present is in San Francisco, father has received a letter. We have heard a lot of wonderful stories about him here, that he has been married three times and divorced as many times. Oh, what a shame and great pity! That man who in our eyes was so gallant! Start well but fizzle out? Yes, he "went up like a sun and down as a pancake!"

One day Messrs Kildahl suggested that we portray ourselves with them. And so we did; we have now received the [cabinet] photographs. They are quite good, and we want to send one home to Grandma, who will probably be very happy to have it.

Monday [2 August 1886]. We were on board all day and received visits, which we had in abundance. Captain Cutler's came here in the company of a family named Kiteley. About Mrs Kiteley there could be a lot to say. She is the cutest, prettiest creature you can ever see and only twenty years old. The family had a small one-year-old son with them. We are invited to them on Thursday. —

Captain Lewis was also here and Mr Kildahl, too, of course. But the strangest of our visitors today was an old acquaintance from home, Manne Gudmundson [of Gefle]. We could hardly believe our eyes when we saw him enter. He has no doubt been through many adventures since the last time we saw him. Anyway, he looked the same as ever. I think it will interest Lully to hear that we've seen him.

Tuesday we were invited with Captain Lewis to Port Ludlow. It was a very nice excursion. In the evening we were again invited to a ball like the previous one but did not go, because we were too tired, and I did not feel well at all.

Wednesday, I lay sick all day, freezing and perspiring by turns, and having a horrible headache. Of course Mr K-l [Kildahl] was here, but I heard his sweet voice only from a distance. How unfortunate!

Thursday, much better! I must absolutely pick up to attend no fewer than *two* very special *parties*, to which we have been invited. First a dinner at Mrs Kiteley's! That was the best thing. They have such a stunningly gorgeous home, and Mrs Kiteley is the cutest little house-wife ever. She and her sister, Miss Cara Carter, are from Honolulu. They told us all sorts of interesting things from there. Among other things, they had been at the Ball that was given by *Vanadis's* officers a couple of years ago. They then danced with *Prince Oscar*.'[86]

---

[86] Natt och Dag, 1887.

That occurred two years previously. The steam frigate HMS VANADIS became famous for its circumnavigation in 1883–1885, a voyage of exploration with the Swedish crown prince OSCAR in the crew, a trip unforgettably documented in some seven hundred excellent photographs by Oscar Ekholm, and resulting in a huge collection of ethnographic museum specimens from all the world. The ship was at OAʻHU, HAWAIʻI, 20 June–10 July 1884. His Majesty KALĀKAUA, the last king of the monarchy of Hawaiʻi (succeeded by Queen Liliʻuokalani, the last sovereign), honoured the ship with His presence.

'At Mrs Gore's—which we could flatter ourselves with since her invitation was personal and especially for our sake—all the youths from Port Gamble were gathered. There was talking, playing, music, and courting!'

22. Return to Port Townsend

'ON FRIDAY, 6 AUGUST 1886, I went to PORT TOWNSEND. There were several incidents on this adventurous voyage. To begin with, when we rowed from the *Atlantic* to the steamboat in the middle of the night, our sole source of light in the darkness, the lantern, fell over, and the dinghy was suddenly ablaze, and we had severe trouble extinguishing the fire. Luckily enough, daddy was fellow-traveller, and he in the event also got me a good seat in the steamboat when we embarked. Instead of reaching Port Townsend at six in the morning, we did not arrive until ten o'clock, as the steamboat ran aground, only a few minutes' drive from Port Gamble after departure. I was both sad and annoyed, but comforted myself and went to sleep.

Mr Rothschild met me upon arrival with horses. I stayed with them until Tuesday afternoon. I actually did not enjoy the stay there very much. Saturday night we were on visit to a very fashionable family, the bank manager Sanders'. They are Germans, as are the Rothschilds, which can be heard by the name, but they speak English. We get a lot of training in English here. The Sanders' had two daughters who treated us to music, but miserably so.

On SUNDAY, Regina and I were both to church service and to Sunday school. The priest spoke a very strange English and had a peculiar dialect, so that I could hardly understand him. At Sunday school I got questions to answer. Regina plays the organ there.

On MONDAY, Port Townsend's streets were filled with people, as a steamer from San Francisco arrived there. We had fun going out and watching the people. I did not meet Manne Gudmundson, although he mentioned that he wanted to search me up there.

I was very happy when it was time to get back on board again. Father was so considerate then, and said that it had been boring for me. Just think that this was possible!

I had a nice return trip. I enjoyed looking at the beautiful beaches. The captain of the steamboat surprised me by mentioning that he was Swedish, he seemed very polite, and I talked to him all the way.

The same day as I got home, Captain Cutlers' left. It is a pity for us that our acquaintance with people we like has hardly begun before we have to part, and probably forever, because our paths go in very different directions. Captain Cutler, who is very funny, said Goodbye to Dad in this exact wording: "I wish you fair and flying wind over the ocean of life, so I wish you good sons-in-law; be kind to your grand-children!" (Figure 14).

FRIDAY THE 20TH OF AUGUST, 1886

My poor little diary has now been completely forgotten for a long time, and it will be with difficulty I can remember how we further on spent our time in Port Gamble. Overall, it has been as before, that we had a lot of fun. We've got new acquaintances, a captain family from the east, New Brunswick. Their name is Copp. The ship's name is "Earl of Granville".[87] We like the lady a lot. They have four children, three boys and a girl. All the children are very handsome, especially little Dolly,

---

[87] *Earl Granville*, ship [barque] of Saint John, New Brunswick, Canada; 1193 tons, Captain William Harvey Copp—as noted in contemporary directories <shipindex.org> and in Lloyd's register—'sailed from Port Townsend to Sydney on November 15th 1886'. Copp was for some time later on also the commander of a notorious English fishing vessel, the schooner *Vancouver Bell* in the North Pacific. Copp is a British

who is indeed very sweet. By the way, they are truly harum-scarum and get us in a flurry when they come over here. Another captain, named Clapp,[88] also has his wife with him. She is about eighteen years old, is charming and lovely, and we have had a very good time in her company. American women usually marry terribly young, often at fifteen to sixteen years of age. Ugh! It's selling oneself too early![89]

LAST SUNDAY we were on an exceedingly agreeable *picnic*. Several families, the Walkers', the Kiteleys', the Whites', the Captain Lewis' and others had been out on excursion for several days and brought tents with them, in which they spent the nights outdoors in the woods by the coast. They have had a very good time, they say, but I think it must have been a bit raw and chilly, especially for Mrs Kiteley's little boy [William]. We are proud now to have made Miss Walker's acquaintance, too. She is a beautiful, stately girl, and looks very classy (almost too much so), but she is also the only child of a multimillionaire, you see!

ON TUESDAY we were invited by Mr Kiteley to a so-called *theatre*, at least alleged to be something of the kind. A lousy drama group has been here with its full troupe of wannabee actors on travel, giving performances for a number of evenings, and their show is all total rubbish.

ON WEDNESDAY Mrs Gore, Miss McPhee and Miss Guptill were here, and later in the evening also some gentlemen, their usual cavaliers. Mr Kildahl, too, of course! He is our daily guest. He is now beginning to show clearly what I previously only had reason to suspect! I have to admit to myself, that I'm a terribly coquettish creature now. Oh, it is a great shame, if I have indeed encouraged him in this courtship, I have of course meant nothing serious by it!

ON THURSDAY, I went over to the "Earl of Granville", I was invited there for dinner, and had quite a good time with Mrs Copp, but after

---

family known since the sixteenth century, renowned in America since the pioneering Massachusetts Bay Company and the American Revolution of 1776.
[88] Unlike his friend Captain Söderström, who persistently and faithfully held to his ship for twenty-four years, Captain G. P. Clapp was the master of several different ships and of various kinds, yet first of all barques, though, from the *Fannie* to the *General Butler*, which he mastered and esteemed as his colleague did.
[89] As observed by the 19-year-old diarist when she saw the old captain on visit.

a while we received word from the *Atlantic,* that we again had received guests, Mrs Kiteley with her little William, and her sister, Miss Carter, had come there to visit us.'

23. On the Sea

'Sunday afternoon the 12th of September, 1886

Even the date, which I have now written down here, shows that I have great reasons to blame myself for my negligence in not having made any notes in my diary in such a horribly long time. When you have postponed something for as long time as I now have done, and to do what should be done, it then becomes twice as difficult to finally make up for it. I have now almost forgotten what happened during the long interval.

There is not much to say about the remaining time we were in Port Gamble. It was as before, we did have a lot of fun, perhaps even more towards the end of the stay, as is often the case. Nevertheless, we finally became very tired by all festivity, hubbubs of laughter and emotional outbursts, and were not at all dissatisfied with being able to set out to sea again, and get back to the offshore peace and quiet. Nonetheless, it was a little sad to be separated from all the new friends there. In particular, it was really sad to part with Mr Kildahl! This made a terrible scene, as I had expected. Yes, it's my own fault, I should have behaved differently from the beginning. However, he will soon forget me! I do not believe so terribly much in the duration of "love". And especially not on the endurance of "boys' love"! It does not endure a blow more than a faded buttercup!

Well, still something more about Port Gamble, now!

For a couple of days, we were busy going on farewell visits, and the day we left, about 30 people came on board to say a final goodbye to us. And we were really overwhelmed with flowers and fruits, etc. So, we can confidently say that we as strangers were treated really kindly.

In Port Townsend we stayed a few days, and were then of course with the Rothschilds. We were there over a Sunday, so we could get to church. We accompanied Mary to the Catholic Church to please her. She had also once accompanied us to the church in Port Gamble. (The whole time we were there in Port Gamble, there was worship only once.) In the afternoon, we also attended Sunday school. The pastor was brilliant. We understood so well his message, and he really spoke to the heart concerning Jesus and heaven, where The Lord had gone in advance to prepare a room for us. I remember how much I felt drawn that day to the eternal abodes of peace, but the tranquillity was somewhat disturbed by a promenade in company with a lot of young people, friends of Regina's. Oh, what a row and racket they made! And yet, they came straightaway from Sunday school!

Monday morning, Emy and Mary went ashore with Dad. I was alone on board, and took the opportunity of asking my God quite zealously for mercy and compassion. Oh, I do not deserve to belong to Him, but still—on behalf of His mercy! At times, I consider myself really bad and execrable; I am indeed so thoughtless and superficial that I get scared. The only good in me is my sincere willingness for betterment, and may God assist me in this.

> A little while with Jesus,
> Oh, how it smooths everything
> and gives to all life a new and bright form.[90]

Yes, only with Jesus is there peace and quiet for a wandering and troubled human heart, ridden by *inquietude*. May I hold on to that!

I have more to add about our stay in Port Townsend!

---

[90] Yet another song by Lina Sandell, whose flow of tender and wistful songs started after she saw his father drown at a boat trip. Sandell's version is, in its turn, a translation (1879) of a hymn called 'A little talk with Jesus', written by the English author Anne Louise Ashley-Greenstreet in 1871.

In the afternoon the same day there were lots of guests here—Regina and her brother, and several girls, friends of Regina. We were not particularly pleased by our visitors then. We who travel "around the world" have a good opportunity to choose our acquaintances, and to choose them well, and as to those girls, we certainly would not want to cherry-pick them. Moreover, I would venture to say, that perhaps they were too—well, how shall I say, perhaps – "sluttish" and slovenly [Swedish: *sjampiga*]! (They had too much of "the American style"!)

The last day, i.e. on the 31st of August, when we left, Mr Rothschild invited us out to go for a ride. We had a very pleasant excursion that day, which extended many miles beyond Port Townsend, all the way passing through beautiful, magnificent forests, dizzy heights and deep valley depressions with the most dazzling views. We came back quite late, so that the steamboat, which was to tow the *Atlantic*, had been waiting for us. Regina and her brother accompanied us when it was time for boarding, and they did not say Goodbye until the anchor was weighed, and we set out to sea.

And now we have been next to two weeks out on the sea again! These have passed so wonderfully fast. We have had good weather and good wind, so we are now already in the tropics. I should not forget to mention that we have a passenger on board, a young Mr Gardiner, a real *Yankee*. Mary looks at him with not very unobtrusive glances. I have not started my English lessons yet, but want to start tomorrow, and intend to be really studious as I was on the last trip.'

24. Back to Australia—Crucial Days

'Melbourne on 20th November, 1886

So here we are again! We arrived here on the 17th of November, after an as happy and pleasant journey, as usually.'

Over half a year, almost seven months had passed since the last time they were in Melbourne—due to the roundtrip across the Pacific Ocean

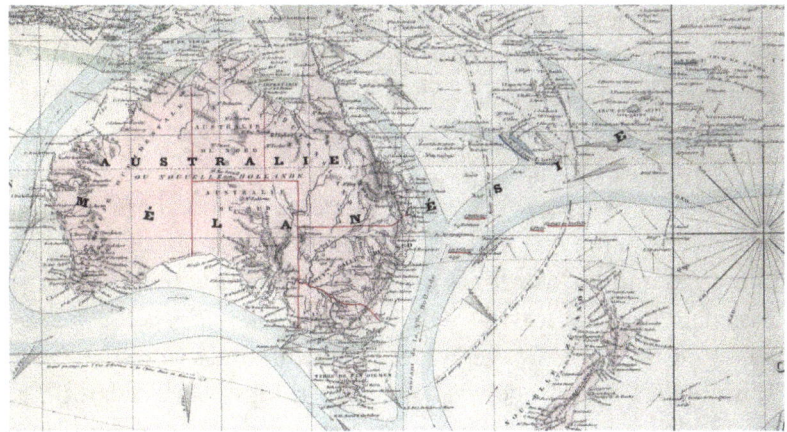

Fig. 15. *The Dufour map of 1863. Passing south of Australia outwards through the Pacific and by northern roundabout homewards was a challenge and called for many stops. The* Atlantic *tried these routes several times, choosing various alternatives.* Private possession.

and the long reloading time in North America. It was on Easter Eve of the 24th of April they sailed from Melbourne. They were back to Melbourne proper on 17 November, 1886, as corroborated by the report of the Melbourne ARGUS *issued on Friday 19 November 1886 about the arrival of the* ATLANTIC *from Port Townsend to* HOBSON'S BAY, *carrying '10,150 pcs (800,000 ft.) lumber, 1,725 bdls pickets, 135,000 laths.' Hobson's Bay is a local government area by the sea in metropolitan Melbourne, Victoria.*

'It was only the last two weeks that we had stormy and nasty weather. We then also had some terrible thunder nights here off the coast. Oh, I'll never forget the impression of them! The sky was like a sea of fire, and the lightnings struck incessantly on all sides of us. I think we have never been so close to death as then; now it was only a warning from God and a proof of His grace that we were saved. Yes, God has luckily brought us through an ordeal and is still as gracious.

During the few days we have been here, we have only been ashore once. We then went to look for the Wohlfahrt girls, but did not succeed in meeting them, because they had moved from the house where they were living when we were here last, and now we know no way to find out where they are staying at present. Mrs Zickerman has travelled home to Sweden. Any other news we do not have about our acquaintances here yet. Oh, I am forgetting: The *Nanna*, with Captain Grubbström and his wife, they are here, and we saw them, albeit only in passing. They are leaving here again in a few days. We were out *shopping* a little, and got several new beautiful things from dad, such as hats, umbrellas, dresses, gowns, and other things. What a darling poor daddy is! He has a lot of troubles and worries, so much to consider, due to the bad times. The general complaint!'

It was increasingly difficult to get good payment for shiploads of wood products in the harbours and find an outlet for all exports—due to rising competition, declining demand and dipping prices. The party was travelling in the middle of a long-term recession, and at this time the choice of building material also gradually turned from wood to stones, bricks, and mortar in cities like Melbourne.

'We are now terribly curious to know where we may go from here. Before, we had half a hope of being able to go home to Europe, and come home for the summer, but now father does not think that is possible. However, we want to put everything in the hands of our dear faithful God. He guides and leads us according to His gracious loving will, and then we can be happy and content, however it may be.

Of course, we have been busy writing letters these days. The letters from home contain only good news. Our little grandmother has just settled down after her move. What a pleasure it will be to see her in her new home, whenever it happens!

Our Mary has now been transported ashore, happy and well. Strangely enough, it's not empty after her at all. That girl was probably kind, but not very nice, and it is perhaps nasty of me that I do not like her more.'

25. South Melbourne Vistas and Manners

'T*HE* 3*RD* D*ECEMBER*, 1886 [F*RIDAY*]

I should keep notes more regularly! I'm so horribly sloppy. A real *slip-shod-Maja*! Nevertheless, I can now appeal to my good intent, because I have indeed opened this poor book several times to write down a few words, but have always been interrupted, because there are other things to do. I have ever so much more trouble now in trying to gather what I have to include, since the last time I kept book. Today I seem to have a good time writing. Emy and I are alone on board, and the weather is particularly suitable for writing—that is, it is rainy days and we lie at anchor.

Speaking of the weather here, it is quite unstable this time of the year: one day, or rather at one moment, there is real tropical heat with wind from inland; and in the next moment the wind turns and blows from the South and it is at once really cold [by the Antarctic southerly wind] on the ship. This makes it very easy to catch a cold here in summer. I have been really ill this last week, yesterday was the worst, I was in bed then and had fever and chills, with fits of shivering. I also feel quite feeble and done up today.

But perhaps the weakness was largely caused by the latest news. We have had a letter from Alma Hummel informing us that she is coming over to us, to stay as long as we are lying here. Let's see how it can be for us in her company! Although she seemed to us extremely lovable and interesting during those few short days in Adelaide, we fear that for so long time now, it will be much constraint and lack of freedom for us, for surely she is a little *goose* with a thousand ideas and peculiarities.

Almost all of last week we were ashore. A couple of days with Mary! She is now back to the household chores at home. She is all in all a really good and kind girl, the apple of her father's eye! No wonder, her poor father just does not have much joy from his other daughter, Julia, who has gone wrong. Something terrible has actually happened with her, unfortunately there is something greatly amiss. She has run away from home and now lives with a girl of bad reputation, and is funded by a gentleman who is supposed to be her fiancé.

The fact is, that she in less than no time got fed up with taking care of her father's household, and escaped to gain her freedom. To tell the truth, she has always been unruly, and a source of trouble and vexation at home. Well, after all, this is just Melbourne life in a flash! That's the way it is.

Her sister Mary's fate and attitude proves the rule. When she accompanied us on the long trip she intimated that she was engaged and maintained this during the voyage, but now ashore her fiancé has disappeared, no one knows where! In spite of everything, Mary takes it easy: "I do not care!" Yes, it was a betrothal only in the usual *South Melbourne manner!*

We have been to dinner at Mr Römcke's. They had a very nice home, a beautiful villa at St Kilda, one of the foremost districts.[91] They appear to be happy, but the wife doubtless seems to be quite "motherly" to him. Basically, she is a very nice and intelligent lady, and pleasant to be in company with.

The beauty of St Kilda is magnificent. There are so many picturesque places here in Melbourne. We were out in the countryside one day, along the shores of Yarra, and we enjoyed the lovely sights.'

The grand Yarra meanders on its way westwards some 240 km through fruitful valleys and flowering meadows to the city of Melbourne. The river debouches after the Docklands and the Southern Warf into Hobsons Bay—where the Atlantic was always sighted by the coast guard—which is part of the large and almost enclosed Port Phillip Bay. This was the natural dock and harbour for the earliest disembarkations on the coast down under, after the passage of the narrow entrance from the Bass Strait, of old remembered as The Rip.

A safe haven and anchorage after hazardous navigation in the waters along the shores of The 37th Parallel South, steering by the stars with the perils of reefs, rocks, and gales, which made sea-approach of Victoria known as The Shipwreck Coast.

Now, in the idyllic atmosphere of visiting friends of the captain in St Kilda, and travelling the magnificent Yarra Valley, the party was relieved from the challenge that the Atlantic perpetually encountered offshore.

---

[91] Saint Kilda is a bay-side suburb in Melbourne, favourite highbrow residence in the late 1800s. Named for the schooner *Lady of St Kilda*, which moored on the beach in the early 1840s, subsequently shipwrecked off Tahiti.

Here this meant crossing the BASS STRAIT *and passing* KING ISLAND *and* TASMANIA, *which it did over and over again in different directions year after year, on its never-ending journey, with the old salt of Gefle at the rudder of the wooden sailing vessel for twenty-three years.*

This explains the number of old friends abroad, colleagues as well as business partners, the skippers often earning their principal income from customary kaplage (= 'coat cloth'), i.e., profit share, the so-called Chapeau de maître, the master's 'hat-money'.

26. The Monster Wave that Struck the *Orient*

'Captain Sätterlund is here now with his barque. It was a real joy to see him again. Several other Swedish ships are here. We were on board of one of them one day at a dinner party, the *Orient*, Captain Ahlgren,[92] CAPTAIN VON SCHÉELE'S SUCCESSOR. Oh, that's awful! The incident with Captain Schéele! He is dead, and such a horrible death! During a severe storm with high seas during the voyage here, he was thrown overboard by the waves.'

IN JUNE *1886, the barque* ORIENT *left Gefle loaded with plank from Stora Kopparberg bound for Australia. In the Indian Ocean, off the Cape on 6 October, the ship's young new master Bror Georg Benjamin von Schéele (born 1855) came up to the secured helmsmen on afterdeck in the tempest, and was suddenly smashed overboard by a colossal wave that pooped over the stern, and the man disappeared in the dark body of water. The steersman took command after the captain had been lost on his first trip with the ship.*

The scene when the captain drowned before the very eyes of his crew, beyond the reach of rescue, was engraved upon the people of the ORIENT *that the Söderströms encountered in Melbourne, in particular the former first mate at the wheel, captain Johan Ahlgren, the successor.*

The accident was a tragedy that deeply affected them all, since the captains Söderström and von Schéele were close friends, and their ships for some time sailed in shipbroker PJ HAEGERSTRAND's *fleet.*

---

[92] Johan Ahlgren, formerly first mate under von Schéele's command, and his successor when the captain had been washed overboard. Ahlgren remained the master for a couple of years [Nordenberg, 2014, 194f.].

Söderström and Ahlgren had much to talk about, and the cabin girls mourned that bright man of letters, a celeb and popular publicist in their hometown.[93] *Moreover, news of this kind filled all seafarers with horror.*

'How awful to be alive one minute and not suspect anything, and to be dead the next minute! Oh, it's important to be prepared! Am I ready to die? I often wish I could come home to heaven, and I long so much to see Jesus up there. But that is in my better moments! Unfortunately, I feel even more often so horribly *indifferent* to everything—*to heaven, yes, to Jesus too!*

> Anxious heart, come with your sorrow and distress,
> Bring your pain, bring your cold and death
> Only to His heart![94]

Yes, I now come to you again, O Jesus, to find warmth and life!

THE 6ᵀᴴ OF DECEMBER, 1886 [MONDAY]

Today, Alma Hummel has arrived. Oh, what a small (yes, small, small!) handy girl she is! I hope we will like each other.

The whereabouts of our friends, the Wohlfahrt girls, we have now finally found out, and they have been on board with us a couple of times; they stayed here on Saturday and Sunday. We've agreed to be together a lot while we're here.

Our old friend, Pastor Carlssen, is here preaching on Sundays as before. He will probably stay here for a long time. It will be a lot of fun to go to church.

---

[93] The author of the pseudonymous recollections of 'John Grogg' from the seas and the harbours, *John Groggs minnen från hafven och hamnarne*. Stockholm: C. A. V. Lundholm, 1889 [Henricson, 2003, 113; Henricson 2004, 152].

[94] From a hymn by the aforementioned author Lina Sandell.

The 23rd [of December, 1886]

The day before Christmas Eve, you always have a lot of little things to do. So also, today. We have been ashore shopping all sorts of small things for dad and for each other, for Alma, for the steward, and others. It has been so much fun to get drawn into the crowds of the Christmas rush of the city, although it can really make the head dizzy. You are dazzled by all the shining splendour in the shop windows; everything is decorated here as at midsummer at home with flowers and leaves. It's all pretty hard for us to imagine, that it really is Christmas. How can it be like Christmas without snow and ice and hoarfrost?

I also do not feel any real Christmas peace either. How much I wish that the same childlike joy at Christmas could be maintained even after getting older! May I search my own heart in prayers to God to find out what it is that makes my mind so foreign to all true, quiet Christmas joy. I'm pretty worried. Jesus, you Son of the Holy God, whose birth we are now given to celebrate, who invite the anxious to come to you, wanting to give them peace. I am coming to you now, Lord Jesus. Oh, may I in faith receive what you offer! Yea, I want to rest in stillness with you.

However, it does not seem that we may celebrate Christmas in the same quiet way as we did last year. All this time before Christmas has also been days of pastime, amusement and entertainment. To record day by day what we have been up to would be too long-winded, indeed. Of course, little Alma has often been tagging around with us all day, but she still has a lot of other acquaintances here in Melbourne, with whom she stays overnight in turns. This last week she has been in WILLIAMSTOWN, but will be back here tomorrow, on Christmas Eve, to celebrate the feast with us. Christmas Eve we will, as I hope, be for ourselves with the crew and with not many guests on visit—and we intend to *eat fish and porridge* as always, and remind ourselves of home. Afterwards, on Christmas Day, we will throw a big party on board the *Atlantic*. All the captains from both the Swedish and Norwegian ships which are here in the harbour, are invited. Besides, all our Melbourne friends and acquaintances, such as the Wohlfahrts', the Grundéns' and others, are coming!'

27. Christmas in Australian Summer

'CHRISTMAS EVE, 24 DECEMBER 1886 [FRIDAY]

Nice and peaceful!

Many Christmas presents! We actually had a real Christmas tree in the stand, donated to us by Captain Sätterlund! Only he and a *Captain Ljungberg* were here!—See the Christmas gifts here: From Dad: a napkin ring of silver each, half a dozen embroidered linen handkerchiefs, half a dozen socks of merino wool; each a pair of silk-stockings, several pairs of gloves, and, in money, one gold pound coin [a Melbourne Mint Sovereign] each! From Alma I received a very beautiful sofa cushion with embroidery on dark red velvet and pink silk; Emy got from her an oil painting (which she had made herself) in a beautiful frame. From Captain Ljungberg, we each received most superb Souper shawls, Emy a light blue, me a cream-coloured. (By another captain we were each given a silver medallion with chain.) In addition, flowers and Christmas cards were endlessly coming in from all directions!

So, after all it ended up real Christmassy in the end! It was probably the Christmas tree, the candles, or also the porridge, that did the thing!

CHRISTMAS DAY

A memorable day!'

28. 'Frankly, nowadays I do no longer dare to be sincere and honest with myself!'

'THURSDAY THE 30TH OF DECEMBER, 1886

An equally memorable and extremely important day! (Frankly, nowadays I do no longer dare to be sincere and honest with myself!)

NEW YEAR'S DAY 1887

I really should compose myself and devote myself to a New Year's consideration and reflection! However, I do not feel particularly apt or suitably tuned for that!

A new year! Oh, what will it bring? I feel like it's going to be an important, crucial year.

> Delight yourself in the LORD,
> and he will give you the desires of your heart.
> Commit your way to the LORD;
> trust in him, and he will act.
> Psalms 37: 4-5.

Look there, a word that holds for building on, for the whole year and for everything to come.'

29. Interlude: The Dream of True Love and the Perfection of Matrimony

The journal momentarily keeps quiet as ever more private and clandestine matters of vital importance are taking place in the life of the diarist, the power and nature of which are shown by the set of poems which was interleaved and added to the notebook. When the narrator becomes silent, a choir of other voices strike up in key. What has been in progress over the preceding year, the writer has vaguely indicated, and the true course of events she will intimate in her recollections much later.

The Diary Poems
She had picked up a folk song abroad, and then she noted it down with a special emphasis in the subtitle:

'Folk song from Thuringia [Thüringen in Germany]
—To be sung when I get a fiancé!—

Me abandon you?
What can you believe of me?
Who, from the bottom of my heart
love you so much,
Beloved! Both 'soul and mind'
I have taken you in,

So that this heart beats
Only for you.

Around you small flowers stand
Blinking with blue eyes
Nodding 'forget-me-not!'
Whispering about hope,
Soon, however, the meadow lies waste,
Do not cry, the abundance
We have in our love,
It shall not die!'

That this selection of reading is concurrent and done in earnest, and not coincidental, is shown by the following extracts from a Swedish poem entitled Allvar written by the clergyman and historian Claës Johan Ljungström (1819–82), which the diarist had read in the monthly Svenska Familj—Journalen of 1876 after an account of Francesca da Rimini's touching love story in Dante's DIVINA COMMEDIA.[95]

'SERIOUSNESS!

When you have found a friend,
Stay in the friendship even when it disappoints you.
As the tree does not suddenly become a mast
But is given annual ring upon annual ring:
Likewise, friendship becomes firmer with each year,
And new ones you established for the day
May be broken tomorrow.
Then nevertheless old acquaintances
Remain that you struck in bygone days.
A friendship that has grown from early days in your heart,
Oh, cherish it well,
that it may still flourish there
when you bid farewell to the world!
Take your friendship seriously!

---

[95] *Svenska Familj-Journalen* (Halmstad: Gernant, 1976) vol. 15, No. 1, p. 25–26.

Are you a man or a woman, and have become attached
to a beloved one, then remain faithful!
A love, which is only a moment's guest
Is unworthy; — One that is for life
And for ever, is from heaven:
It is an angel, whose robe of clear light
No betrayal, no infidelity can touch,
Not even the slightest breeze.
What you promised about love
and promised about faithfulness,
Keep that promise until the heart gets cold.
On your way henceforward,
roses will then constantly grow
and flourish for you everywhere.
Take love seriously!

Are you still in the carefree spring of youth,
Or have reached the height of life,
Or stand near the grave with whitening hair,
Be you woeful or joyous,
Look up at the sky with the gaze of love,
Love God both in life and in death;
He loved you first, and however everything went for you
Everything went to you as He commanded,
He is the Father, you the child; and His wisdom vouchsafes
What He finds best for you, and
When you believe Him to be farthest
He is closest,
For He has attached you to His heart.
Love God, but seriously!'

The following two poems she noted down in English.
Her longing for being in safety, to be cherished and carefully treated by a beloved, as well as by her accompanying sister and all good forces, is evident from the familiar English nursery rhyme she took down from memory in this travel diary on the sea. A most striking coincidental reference in this rhyme is the remembrance of the

girl's dead mother, the loss that made them end up in a sailing ship cruising the world.

The old original, slightly different, was written or adapted by Dexter Smith and set to music by C. A. White and published by White, Smith & Perry, Boston: 1870, and was widely known those days.

'Put Me In My Little Bed

> Birdie, I am tired now,
> I do not care to hear you sing;
> You sang your happy song the whole day,
> So, put your head beneath your wing;
> Come, sister, come, kiss me good night
> 'fore I my evening's prayer said
> For I am tired and sleepy too,
> So put me in my little bed!
>
> Birdie, what did mother say,
> When she was taken to heaven away?
> She told me always to be good
> And never, never go astray,
> Come sister, come, kiss me goodnight
> 'fore I my evening's prayer said
> For I am tired and sleepy too,
> So put me in my little bed!'

The larger context is clarified in the third verse, which she did not write down or remember:

> Dear sister, come and hear my pray'r,
> Now ere I lay me down to sleep,
> Within my Heav'nly Father's care,
> While angels bright their vigils keep;
> And let me ask of Him above,
> To keep my soul in paths of right,
> Oh! let me thank Him for His love,
> Ere I shall say my last 'good night'.

HOME, SWEET HOME

This is a verse adapted from the American dramatist John Howard Payne's opera CLARI, OR THE MAID OF MILAN *(1823), which became an enormously popular ballad with a melody composed by Sir Henry Bishop. It was sung in North America up to the Civil War and beyond, and was used and quoted by the Swedish composer Franz Berwald in 1827. In this case, too, Mia is writing down her (a bit differing) text from memory, which is probably the words the homesick sisters sang on board with the crew.*

> 'Mid pleasures and places though we may roam
> Be it ever so humble there is no place like home,
> A charm from the skies seems to hallow us there
> Which seek through the world, is ne'er met with elsewhere
> Home, home, sweet home!
> There is no place like home!
>
> In exile from home splendour dazzles in vain
> Oh, give me my lovely thatched cottage again,
> The birds singing – gaily that came at my call
> Give me these and the peace of mind dearer than all.
> Home, sweet, sweet home
> There's no place like home!'

Maria Söderström has many doubts about how to become a good and able wife in the future, and she ponders upon how to live happily and rightly by rectification. Among her reminders was the following time-honoured and demanding ideal, which is also part of the young women's heritage, although hard to live up to. Maybe foreign to those busy hands on a ship's deck in the world down under who were necking in the moonlight and aiming for the stars—and yet noted down in the reflective and increasingly animated journal of travel.

'HOW A YOUNG PERSON SHOULD BEHAVE!

(Occasional poem written by an eighteenth century Swedish priest.)

> If girls understood
> What is good for them

And redound to glory,
Then they would learn
To read and write
And be polite
To sew and to twist
To weave and spin,
To whip their cake
To brew and bake
To cook and fry
Instead of play.
Then they would appeal,
Get ornaments and jewellery
Grooms and men,
For bachelors know
That a husband is blessed
If the wife is nice.
But inertia and laziness
And cheekiness and conceitedness
Brings poverty and whining to the household.
Morning and noon
Midnight and evening
Remember this, Mam'selle!!'

The following folk song or ballad found in the notebook was first published in a popular so-called 'shilling print' or chapbook. It was adopted by the Swedish poet Gustaf Kahlson (b. 1833) and adapted and set to music by the famous composer Adolf Fredrik Lindblad (1801–1878) and it was read or sung as a hymn. The theme obviously appealed to the brooding and wandering girl who was constantly faced by moral issues and occupied with self-reproach. The symbolism of the sea is unmistakable, too.

'Do Not Judge! (by Lindblad)

Not of bright colours the lily flaunts,
Yet its inner splendour is probably seen,
Love in the eyes of many sparkle
Though you cannot grasp its value.

The cliff only shows its hard surface
But underneath the veins of ore go;
Storms often break the surface of the sea
But in the deep it is calm all the same.

Many have a sharp and bitter tongue
But in the heart, however, gentleness can dwell.
Many feel compelled to joke, even sing
Though in the heart the thorns of sorrow grow.
Many of the poor are apparently rich
Many rich people are burdened by afflictions.
Many the son of happiness,
Happiness betrays.
Many mourn deeply
even though their mouths smile.
Alas, then judge no human harshly
Although you think that you see her shortcomings.
Appearances are deceptive, can confuse:
Man's vision is shallow.
First try yourself to be pure and faultless
Do not blame, let injustice be forgotten
And you will in due course experience
That the one who does not judge
will not be judged either!'[96]

The next verse stems from DIKTER FRÅN FRÄMMANDE LÄNDER, *'a collection of poems from abroad,' translated by Carl Rupert Nyblom, member of the Swedish Academy, and published in Upsala in 1876. The author of the poem was the German poet* FRIEDRICH RÜCKERT (1789–1866) *whose romantic lyrical works were well-known in the musical settings of a number of famous composers—ranging from Franz Schubert and Johannes Brahms to Robert and Clara Schuman. Pieces of this kind were performed in the musical Söderström family and even performed and sang on the* ATLANTIC.

---

[96] Besides a deliberation of Matthew 7 this was Maria's former home-room teacher Klara Johansson's moral philosophy in a nutshell.

'The Reunion

> Now he has come on the wings of storm,
> My heart's greeting to him is ringing!
> Who would have guessed that he from his path
> The shy flower would detect?
>
> Yes, he has come on the wings of storm
> My poor heart subdues his gaze
> My heart he has stolen, have I taken his?
> Alas, nay, they beat with one another.
>
> Yes, he has come on the wings of the storm
> And the joys of spring he brings to my soul,
> My autumn he chases, yet I do not complain
> For mine he is all days to come!'

As a sequel, this theme is developed in another German poem she appreciates and which in ecstatic devotional terms expresses the heavenly union in the kiss, reminding of the Song of Songs and the unio mystica of St. John of the Cross or Santa Teresa de Avila. She avoids translating it from the German, as if she were guarding a well-preserved secret, in a tongue not known by all others, and stored in her private book of notes. Now the mood is passionate with a vengeance.

'Der Kuss

> Glühend, wenn ich könnte, wenn ich könnte auf die Lippen gäb'ich,
> Theurer, einen Kuß dir, einen Süßen Liebes Kuß.
> Schmeichelnd sagt' ich Dir manch tändelnd Wort der Lieb' und Zärtlichkeit.
> Manch tändelnd Wort der Lieb' und Zärtlichkeit.
> Sitzend Dir zur Seite, Dir zur Seite
> Spräch' ich immer mehr unter Lachen, unter Lachen
> Tausend Scherze Dir ins Ohr.
>
> Schlaugen hört' ich laut dein Herz, dein Herz
> an meiner klappenden Brust.

Mich reizt nicht Schmuck, nicht Goldgeschmeide
Nicht sehnt mein Herz sich nach solchem Tand
Ein Blick von Dir ist meine Wonne, meine Wonne.
Und ein Kuß, ja, ein Kuß ist meine Lust, meine Lust.
Komm' o komm', was soll das Zaudern?
Komm' zu mir, o komm' zu mir,
Sieh' ich harre, o, komm' zu mir!
Ach, komm' zu mir, lass an deine Brust in der Lieb' Entzücken,
von Dir umfangen, selig mich ruhen
Glühend wenn ich könnte, wenn ich könnte auf die Lippen gäb'ich
Theurer, einen Kuß Dir, einen süßen Liebeskuß.
Ach, komm', ach komm', ich harre Dein, ach komm', o komm'!'

Translation:

THE KISS

Glowing, if I could, if I could, I would give you,
Dearest, a kiss with my lips, an ardent kiss of love
Flatteringly tell you words of love and tenderness
Many a sweet word of love and tenderness
Sitting by your side, by your side
I would be talking more and more while laughing, laughing,
whispering a thousand jokes in your ear.
I hear your heart, your heart, beating loudly
against my pounding chest.
I am not attracted by jewellery, not gold,
my heart does not long for such trinkets:
A look from you is my delight, and a kiss, yes,
a kiss is my pleasure, my Lust.
Come 'o come', why should we hesitate?
Come to me, o come to me,
see I'm waiting, o, come to me!
Oh, come to me,
let on your breast in love's delight,
let me be embraced by you, and to rest in bliss.
Glowing, if I could, I would kiss you dearly,

kiss you dearly, give you a sweet kiss of love.
Oh, come on, come on,
I wait for you, come! come!

einen süßen Liebeskuß.
Ach, komm', ach komm',
ich harre Dein, ach komm', o komm'!

Those were plain words filling a lacuna in the record where the memoire remained silent. The writer's choice of poems is tell-tale!

Finally, the 'Björkens visa' (The Birch's Song), written by Finland's famous poet Zacharias Topelius *at Ruovesi in 1853, discloses to the dreamer what can happen in the moonshine—as in Gui de Maupassant's* Clair de la lune *of 1883. Again, she is quoting from memory a text she knows by heart.*

'The Birch's Song. (Topelius)

The birch tells what can happen in the moonlight.

By the clear strip
Of a flowery shore
A birch sometimes sang
Its green ballads
And at a branch I listened
And once I heard him sing like this:

Ah, I know, I know
Many a secret
Many a girl wept
Under my branches,
Many a boy watched
Here so warm in mind
The wave in the blue bay.

The clear moon shone
On my green branch

And then came one of those
Who cut names in the bark
And there was only one name
And he kissed it
And this has nobody, nobody seen.

Next night there came
a girl who
often looked around
And cut name in the bark
And there was only one
And she kissed it
Which no one, no one has seen.

The next evening
the moon shone on my branch
Then came one again
And then returned the other
Silent as a haven
Searching for the friend's name
In my faithful white arms.

And my white trunk
stood quite grave
And then it was clear
What they both had written
And that they probably would notice
What was concealed in the forest
And I looked at it and smiled.

Then it so happened
that the two overlooked it
And did not kiss the tree
Little friends, how easy,
Does this not happen
When no one, no one has seen.

But as of a sudden there was a light
Over branches and foliage
And a whistling over leaves and top
When a star's tear
Was shed in the trace of the wind
And fell to the earth's short spring.'

The Diary Continues:

30. Romance: Recollections of Love and Longing—*Heading for Micronesia*

'27 March 1887

A long leap in time!

I have a lot to look back on. Yes, how much has not happened in all this time of about three months!

We have long since left the Melbourne dust and the close atmosphere of the city.

Since the 24th of February, *we are on a trip to Malden Island (Figure 17).*

It is with great longing and impatience that we anticipate our arrival there, and may our stay there be quite short; for from there—let us embark on our journey home!

I rejoice when I write these words down, Oh, in just a few months we will, if all goes well and we arrive safely and sound, be at home in our dear Sweden. How exceedingly joyful it will be! What reunions will not then take place! I now have twice as great reasons to long for it, for then I hope to see and embrace My Own John, My Dear! Yes, Now I Have One, and with all my heart I am attached to him, and the certainty that I possess his love makes me indescribably happy, and I feel grateful to God, who has made me so happy.

Our stay in Melbourne, Oh, how important it became, especially for me! A great event occurred, namely that we happened upon a certain Captain John Ljungberg (Figure 16), who happened to *fall in love* with me, and finally I also began to understand how deeply I loved him. He really took me by storm, how could I not love him? I probably tried, to the very last, to deny it to myself, too, but then finally, when he one day asked me if I would like to be his little wife, I could not help but answer YES. So, we became engaged, and experienced a few weeks beyond all description of happiness—Oh, I would like to live these weeks a thousand times again!

But then came the farewell!—Oh, what a goodbye! Nevertheless, hope loomed encouragingly in the distance, the hope of such an even happier reunion! Oh, how parting and divorce are difficult! Yet, without them one would not know anything about how sweet it is, when friends, united with the bonds of the heart, clearly feel that because they have left their hearts with each other, no distances can separate them; their thoughts, feelings, sympathies and desires are the same, they are apparently "one heart and soul". The very longing and loss, no matter how painful it feels, make them feel close to one another, for in them they meet, and above all in prayer before God. Yes, it feels very good to pray for my own darling, that God's fatherly eye may follow him in all his ways, and that God in his great love wants to let us meet again in due course of time. In that hope, and in the happy memories of what has happened, I want to live happy and content and try to drive away as much as possible the sad thoughts that the loss and longing incessantly want to evoke.

In my diary, of course, should be noted the memorable days, when these important events occurred. All so much the better, that they can never be as well engraved here, as they are in my memory!

However, it was ON THE 13TH OF JANUARY, aboard the *Gevalia*, with our sweet, nice and much cherished uncle Sjöström, that the declaration of love took place!'

Captain Gustaf Adolf Sjöström, commander of the ship. The GEVALIA *was a three-masted, carvel-built barque of oak and pinewood, sheathed with copper like the* ATLANTIC. *It left the stocks in 1874, and plied the Australia route with success, except for a collision with a steamer on its way to Adelaide in 1876. It was a merchant vessel in Swedish-Norwegian timber and woodware export, as was Captain Söderström's*

ship, and the two masters knew each so well, that the spry galley girls called the man 'uncle'.

'The moment when he wooed, certainly did not come completely unexpected. I had good reasons to entertain some happy suspicions, especially after the unforgettable Christmas days!'

THE SEAMAN JOHN LJUNGBERG *went down on his knee to propose. As it happened, then the gregarious ship's pig, which this nice day in the harbour roamed freely aboard on its regular, took a positive interest in the genuflecting man, and poked his snout into the matter and nudged the kneeling man in the back, and in this piquant rustic manner set his friendly mark on the romantic scene of the wooing. The suitor stood his ground, though, and all three were happy. Like other ships, the* ATLANTIC *had a pigsty, and the young lady was as familiar with domestic animals on board as her attentive admirer was.*

These friendly ship pigs had accompanied sailors throughout history and were often very cool, and faced the hardships on sea with equanimity. It is said about the Greek philosopher Proclus, that he at the heaving of the sea pointed to the pig on board, which set a good example by unperturbed eating his food in peace and quiet while the crew was upset and in uproar.

'THE DAY AFTERWARDS, *on Friday, the 14th of January*, WE EXCHANGED RINGS. Then the union was sealed, and "the golden bond" put on! Oh, it's certainly not heavy to carry! As a gift from my John, I then received an extraordinarily beautiful gold watch. On the subsequent Sunday, we had a small engagement party here on board: the guests were Captain Sjöström, Mr and Mrs Römcke, Hermann Schultz,[97] and Captain

---

[97] The Schultz family arrived in Australia aboard a German emigrant ship, the barque *Dockenhuden*, in April 1849, from Wellersdorf, Silesia, and settled down in Mill Park, a suburb of Melbourne, Victoria.

Fig. 16. *Captain John Ljungberg, Maria's great love.* Private photograph. The Maria L. Hallengren Collection.

Wahlström.[98] These were the ones which we most went out with in Melbourne.'

JOHAN 'JOHN' ERIK LJUNGBERG *was born in 1852 and thus considerably older than his teenage fiancée. She was 19, he was 34. This difference in age (fifteen years) between spouses was very common and did not attract attention. Many of her father's old colleagues among captains had considerably younger wives, so this was regarded as normal.*

When they met in Melbourne, John was the acting master of the barque GURLI, *which left the dock in Gothenburg in 1879, and plied the Gothenburg–Adelaide route for ship owner G. D.* KENNEDY *of the city. In*

---

[98] May refer to Axel Gustaf Samuel Wahlström of Gothenburg (1847–1923), married to Alice Maud Scaton from the United Kingdom.

its time, this barque was considered one of the most beautiful ships that had ever been built at the shipyard.

He had been a sailor all his life. He was enrolled in the Westervik mercantile marine office (on the Swedish east coast by the Baltic Sea) in 1867, and when 15, joined the team of a captain Sjölander as apprentice and hand. The offspring of a family of seamen, he was the son of Captain Sven Ljungberg, and grandson of yacht master and merchantman captain Johan Ljungberg, born in 1785.

John had a dimple in his chin, and dark blue eyes. The wife of a Danish-Norwegian Seacaptain remembered him as 'Den dejlige Kaptain med de nydelige øjne' (Danish): 'The lovely Captain with the gorgeous eyes.' (Figure 16).

'After the engagement party, it was not many days that my John and I owned each other, because already *on the 24th of January*, he had to travel.[99] Oh, what a terrible day, when the *Gurli* weighed anchor and set out to sea! I had promised my darling to show courage, so as not to make it too heavy for him, but how did it go? Oh, I know he saw how heart-rending it was *for me*! But now *quiet* about these dark memories from the past! It was lived through, and with God's help I did not perish, no matter how horribly dark it seemed then!

However, the time we stayed in Melbourne afterwards was very drab and sad ever since John left, yet not only for that reason, but also because our little beloved father became ill, really badly ill, so that he even had to go to a hospital to get proper medical care. It was a time of worry and anxiety, especially as the doctors claimed that it was a kind of typhoid fever, which could become dangerous and long-lasting. Poor father was very sad and worried that he would have to delay the ship in this way, which well made his illness even worse. He finally recovered enough to leave the hospital, and we were then accommodated for about a week with a family in CARLTON, one of Melbourne's most

---

[99] The *Gurli* left for Europe with a cargo of wheat [Kjellgren, 2020].

beautiful and healthy suburbs, where Dad was prescribed to enjoy fresh air, and take good care of himself to recuperate, so there we went.[100]

In fact, his recovery went very fast then, so that he could start thinking about the departure, and as already mentioned, we *left Melbourne on the 24th of February* [1887].

The only joy and encouragement during those sad weeks was my John's heartfelt letters, of which I received quite a few from Port Pirie, South Australia, where he had gone to load wheat for Europe. Consequently, he is coming home for the summer. Oh, how we also wanted and hoped to get a similar shipping directly home, so that we could come home at the same time, but this did not happen. The *Gurli* now gets such an enormous long lead, and will probably come home a couple of months before us, and possibly it could happen that he had already had time to go out again, before we came back, and then we would not be allowed to meet at all! Oh, this would be terribly distressing, but I have earnestly asked God to be able to be pleased with His will, and, in addition, John and I have encouraged each other to do so.'

ITINERARY: Departure from Melbourne on 24 February 1887, arrival to Malden Island (Figure 17) on 5 April, after forty-one days of travel. 7,368 km or 3,979 nm had the course been straight. But they were following a curvature drawn on the table from Tasmania and New Zealand NE via New Caledonia, then north of Fiji and by Tuvalu, initially with trade winds SE 15 mph, heading in the direction of the Phoenix and Line Islands. They made their customary five knots plus on the average, although they got into belts of NW winds and calms, and passed numerous islands and encountered many ships on the trip, day by night, week after week, approaching the equator in mid-ocean. They had experienced that going in the opposite direction could be trickier, depending on season, weather, and port of departure. In the long roundtrip to North America, with the

---

[100] They stayed at Carlton Gardens, where they walked in the park and were amazed by the magnificent *Royal Exhibition Building*, built 1879–1880, then open to the public, and still standing. In 1887 it was further embellished to house the Centennial International Exhibition of 1888 (Figure 23).

reversed set of currents and trade wind systems in the northern hemisphere, they had passed New Guinea, Banda Sea, Spice Islands (Maluku), and Java on returning to Australia from the north, encircling the continent anticlockwise, and made halts on the way (Figure 15).

31. Arrival to Malden Island

'On the 5th of April, 1887, *we arrived to* Malden Island. We've had a pretty quick and good trip, but badly enough we still feel unhappy and discontent. The discouragement and general frustration here on board is due to the fact that we arrived two days later than expected, that is: two days after the arrival of the ship *Chili*, which departed from Melbourne at the same time as us, and went alongside at first.

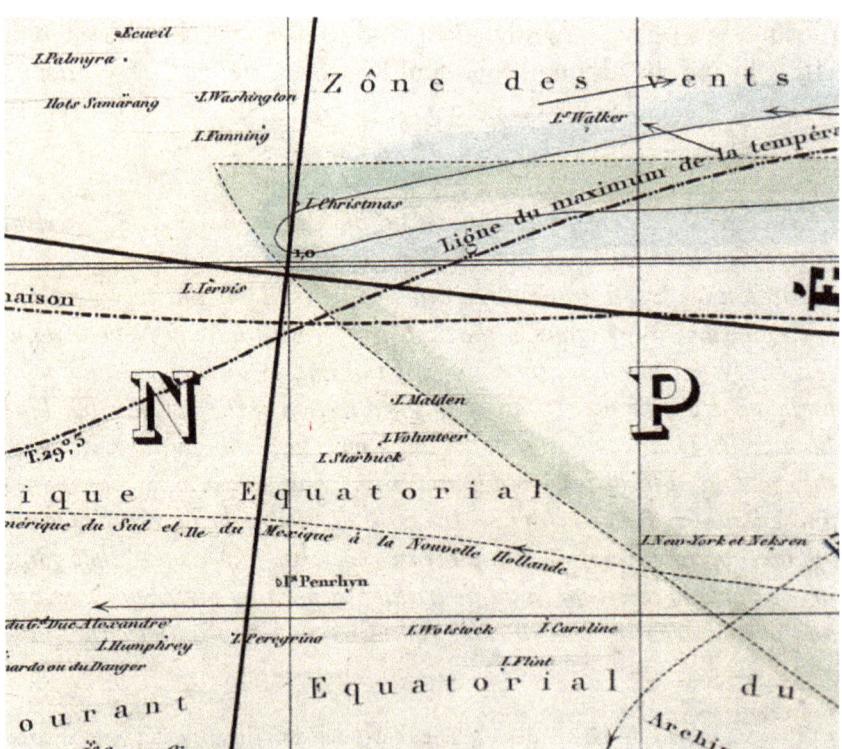

Fig. 17. *To the crew of the Atlantic, the important destination Malden Island was in the middle of nowhere. The Dufour map of 1863.* Author's copy.

A lot depends on this circumstance, because annoyingly enough no more than one single ship can be loaded here at a time in the harbour, and we now have to stay here for so much longer time to wait. I have tried saying "God's will be done," but reluctantly. However, we are not so unlucky as is the *Daphne* astern, yet another Gefle ship, navigated by a Captain Pettersson, now on the waters behind us, which now has to lie and wait offshore for an enormously long time in the file of ships.'

Curiously enough, the strange throng of vessels in this special sailing race, all eagerly scrambling for precedence in pursuit of valuable primary produce of great second-hand market value, were all of the same origin, age, and type, and belonged to the same fleet of merchantmen. Moreover, they all had the same peculiar aim.

At this moment, as seen from ashore or from the horizon, they formed a magnificent sight of tall ships lining up in the sea heading for the small wharf at the odd British possession in Micronesia.

It was a treasure hunt of sorts, in a time of downgraded trade conditions, and falling prices in timber-dealing, which made the captains from the forests of the far north navigate in these distant waters—as a last resort becoming traders in much sought-after manure in an era of poor crops and mass migration. Mining petrified droppings of birds was coining money for the retailers.

The Line-up:

CHILI, *a barque, carvel-built of oak and pine wood and coppered at O. A. Brodin's shipyard in Gefle in 1877. Interior with carved lions, turned oak banisters, and gangway ladders and stairs of teakwood and eucalyptus. Armed with cannons against the pirates. Known not to leak more than a glass of water on Atlantic passage. Captain* ANDERS JOHAN SÅGSTRÖM *commander and at the helm since 1879, when he had Rickard Garberg and Georg von Schéele as mates, both friends of captain Söderström and referred to elsewhere in this book. Destined for Australia already at its maiden voyage, regularly loaded with timber and wood products in Swedish export. Sailed for Peru in the guano trade until this business was stopped in the 1880s by the war with Chile. 671.15 register tons, 157.6 feet, draught 18.9 ft, with a crew of eighteen. Signal: HNSV. Finally foundered and lost in 1902.*

The barque ATLANTIC, *988.64 register tons, 187 feet, draught 22.20 ft, built in 1876, with a crew of nineteen, including the female 'cabin boys'. Signal: HNJB. In 1879 the* ATLANTIC *had collided at* LIZARD POINT *with the British brig* ANTHES, *which sank and lost half of its crew. The* ATLANTIC *was at last shipwrecked in 1911. Captain Axel Söderström's elder brother Carl Gustaf was a sailor who had died in the* NORTH SEA *in 1856, and when their father captain Jonas Reinhold gave Axel the dead sibling's clothes, he also inherited the career of his forebears and went to sea.*

The barque DAPHNE, *built the same year as the* CHILI *and at the same shipyard as the two others. 157.6 ft, draught 18.8 ft, 714.45 register tons. Bound for Australia from the very beginning, but it also chartered grain and coal between European harbours and to South America. Master:* JOHAN PETTERSSON, *with a crew of fifteen. The* DAPHNE *was accident-prone and had been struck by a devasting tidal wave at Bordeaux in 1882 and then collided with other ships. Next year, then under the command of Pettersson, the ship had on way to* VALPARAÍSO *been hit by heavy sea as early as at Dogger Bank and put into port of refuge at Cove's harbour with broken mainmast and helm.*[101]

These North Sea waters in the Atlantic had seen many causalities in the early 1880s. For example, in 1881 Cove was disproportionately hard hit by the EAST COAST FISHING DISASTER, *losing three out of four boats. That disaster was a severe windstorm that struck the south-eastern coast of Scotland, United Kingdom, specifically Berwickshire, on 14 October 1881. 189* FISHERMEN, *most of whom were from the village of Eyemouth, were drowned. It became known as the* BLACK FRIDAY, *and this weekday thence considered evermore ill-fated by seamen. This horror was preceded by the* MORAY FIRTH FISHING DISASTER *of the August 1848 storm, one of the worst fishing disasters in maritime history and never forgotten either.*

Nevertheless, the DAPHNE *was not capsized and wrecked until 1891, and in April of 1887, Captain Pettersson and his men were thus last in the road in waiting for the white gold at* MALDEN ISLAND *in the Pacific:* GUANO, *excrements of seafowl, a fertilizer rich in nitrogen, phosphate, and potassium, and in this area primarily exploited by Australian interests. See there the reason for the line of Swedish ships making a detour from the*

---

[101] Ship descriptions in Nordenberg, 2014; Henricson, 2003 and 2004.

Melbourne route! The demand and the purchaser were Australian, and part of the British Commonwealth.

The cause of many wars and occupations overseas, particularly in the Americas, the deposits of guano beyond the Chile monopoly were becoming increasingly important, even more since the raw material was also used for gunpowder and explosives in the late 1800s.[102]

MALDEN ISLAND is an atoll in the Line Islands with its lagoon enclosed by land. It was (re)discovered in 1825 by Lord Byron (a cousin of the poet), commanding the British HMS Blonde. Populated in prehistoric times, the island in 1887 was solely inhabited by low-paid miners, and the harbour workers who operated the single quay and loading platform at the embankment. The produce was cut by manual labour with picks, brooms, and shovels, and after sifting accumulated in the store house heaps or put in sacks for transportation.

It was a hard and dust-laden loading with tube, manual crane with hemp ropes, lighter, a smelly unhealthy work and a real slow and strenuous business. The captains from Gefle were satisfied if they could fill all holds with the heavy mineral cargo down to deadweight, because they were sailing for deep-water ports, though had to keep draft, displacement and burden tonnage in mind to keep clear of perils and harbour dues.

Those in the crew who were not busy with the cargo, stayed ashore, surrounded by flights of lively gannets and frigate birds, or indoors well off the cargo-hatch and the stowage space, while the heavy thuds and jerks of the hull during loading went on in an air of stone dust day after day.

So much the more overwhelming, and breath-taking in another sense, was the roundtrip proper, since they passed on voyage a less known part of the South Seas, and occasionally anchored to Samoa, Fiji, and New Caledonia on the route.

---

[102] Dung was the unequalled natural manure until it was driven out of competition when ammonia could be synthesized from nitrogen and hydrogen in the air, a discovery that made artificial fertilizers streamline agriculture worldwide and was awarded with the Nobel Prize for chemistry in 1918—an invention with vast environmental consequences. Finally, the British thermonuclear weapons tests in the 1950s forever sealed the fate of Malden Island mining.

'On the 14th of May 1887 we left Malden Island, and are now, thank God, on our way home. A long journey lies ahead of us, and many dangers could meet us, but "we have a God, a God who helps us, and the Lord, the Lord, who saves from death." Yes, God is our mighty protector, and "He hath hitherto helped, and henceforth he will help." "God is faithful and a rock forever."

Oh, what faithful promises and assurances! I want to trust them confidently, and day by day during the journey through earnest prayer call upon me and all my loved ones the blessing they bring. Alas, for my own beloved bosom friend rise daily and momentarily the warmest, most heartfelt prayers to God. He is also still out "on the deceptive sea", but soon with God's help the time is right, when I can imagine him in port. I wonder so much what God's will may be with us. But keep silent my self-willed heart and bow down gratefully and humbly to our God. Bide the time with patience. How much do I not have to be happy and grateful for? First of all: I am a child of God. I have now in a special way felt convinced of that. Yes, in spite of my sins, in spite of all my unworthiness, in spite of all my unbelief and doubt that I so often let prevail, it is true that Jesus is my Saviour because of His unfathomable love and mercy. "To this day I have had the help that comes from God (it's God's grace), and so I stand here". (Acts 26:22.) He who has begun a good work in you, he will also finish it until the day of Jesus Christ.

Oh, may my own sweetheart John also experience this, which I now know so deeply through the grace of God, namely, that the foundation of all true happiness is to be found in God and only in God. What is all earthly happiness without God? Only emptiness and transience. And what follows then? Well, temporary divorce from God can only lead to eternal divorce from Him. Oh, God, for that matter awaken in my beloved John a sincere need of heart for you, for his Saviour. For the sake of your great love I call upon you for it, and for the sake of the holy blood of Jesus that has also flowed for him and me, oh, let it not be wasted for us.

A firm agreement between my John and me is that for the rest of our lives we will join our God. I was then reminded and received again the precious words of remembrance and promise after my confirmation: "I will remember my covenant with thee in the days of thy youth, and I will establish unto thee an everlasting covenant." *Ezekiel* 16:60.

Oh yes, may that covenant become more and more fervent and firmly established year by year. What a happy future then lies before us with God as our friend and helper in everything! Jesus is the first man in our covenant! With such a bond of union between us, we also become more and more deeply united with each other. Jesus makes our covenant infallible, indissoluble. How jubilantly happy these thoughts make me! It's a real relief to be able to write them down, because I do not have my dearest friend here to let him know. (The heart sometimes becomes too overcrowded, overwhelmed, and therefore, dear little diary of confidence, I turn to you.) I sometimes think that I have such great reasons to be merry and exultant that I could scream loudly with joy. Oh, that I own my John's love, is not that great and wonderful? Oh, if that happiness were to be taken away from me, would I be able to bear it?

It seems to me at least now as if all the happiness and joy in life would then be over for me. But that can and must not happen. No, I trust my own darling and believe and know that he can never stop loving me. We have looked into each other's eyes, and read that we belong to each other for the rest of our lives. Oh, nothing, nothing, may separate our hearts. "We want to be joyful and happy in the knowledge that we have each other's love, which neither time nor distance can weaken," he wrote in his farewell letter. No, our love cannot be weakened, with yearning and loss it grows more and more and becomes deeper, firmer, truer by the sense of vacancy and need.'

32. Remembrances of Malden Island

'Since I have now for a good while been up in the clouds "soaring in blue space," i.e., in the memory of the "expanse of blue love," it might not be too unreasonable now to "float" back down to the more prosaic globe, that is, to MALDEN ISLAND, and collect some memories from there. In all probability we have now seen the last glimpse of it forever.

We wished terribly much from the first moment we got there, to get out of there again as soon as possible, for it did not look very inviting. But how great was our surprise that on the contrary and quite unexpected, the time of approximately five weeks we spent there was really nice. There was a lot that interested us there very much, especially everything that concerned the native population of the region. It was great fun to talk to them and to get to know them, and learn about their customs and manners. Some of them were truly intelligent, most of them spoke quite good English, which they had learnt from the missionaries on their own islands. These natives were namely living at Malden Island only periodically, employed by the Melbourne Company, which owns the island, as workers at the guano fields and for the loading of ships.

As a reminder, I would also like to note the name of their *native islands* here: Aititoki and Nuia. We have heard so much about these two islands lately, and have heard them described as so inviting and glorious, that we can only wish to get there to enjoy some of its glories. It is also possible that we will pass one of them now during the journey, and then it is certain that we will exchange for ourselves well with fruit: oranges, pineapples, bananas, coconuts, which all grow there abundantly. A couple of weeks ago, a Melbourne schooner from Aititoki came to Malden Island, and then we got a lot of fruit. Malden Island, on the other hand, is bare, and not that well favoured by nature. At least it is not endowed with anything but guano!'

Excursus On Aitutaki and Nuie: The spelling of these place names in the diary is phonetic or rather onomatopoetic, and not drawn from nautical chart or map. The toponyms are heard, not read, and answer to the perceived pronunciation. The languages of the Pacific were spoken, rarely written, before late colonization, and the cultures had been verbal, sensual, and practical—founded on memory and oral tradition. Only the easternmost part of Polynesia produced a written language of its own, the Rapa Nui of the Easter Island.

The two islands here referred to, are today commonly known as Niue and Aitutaki, but the articulation differs with speakers depending on origin.

The distance between these islands is vast—over a thousand kilometres—and they belong to different cultures in the South Seas. The workers employed by Australian companies came from afar and spoke different Polynesian tongues, whereas the islanders in the Malden Island region spoke Micronesian languages.

The distances of the ocean are well illustrated by the fact that NUIE is found in the middle of the triangle between Tonga, Samoa, and the Cook Islands—to which AITUTAKI belongs. The employees were far away from home, indeed, as were the Australians and the Swedes. The people from Aitutaki, those workers nearest native land, had come some two thousand kilometres by sea to the job at the mine on Malden Island in present-day Kiribati, in the middle of nowhere about the 4th parallel South.

Like Malden Island, the smaller but of old densely populated AITUTAKI north of Rarotonga is a triangular-shaped atoll with lagoon and reefs, with its firmest ground and landing place on the westside. Volcanic soil in the northern part provides the tropical fruits and vegetables mentioned in the diary and have made the island inhabitable. The rumours of its beauty among sailors, reflected in the journal, in course of time made it a tourist paradise like Christmas Island (Kiritimati) and termed 'the world's most beautiful island'. The first Europeans to arrive in Aitutaki, were actually CAPTAIN JOHN BLIGH and the crew of the HMS BOUNTY in 1789, brought up earlier in this book.

The huge NIUE, on the other hand, named SAVAGE ISLAND by James Cook, who was refused admission by hostile residents defending their territory, is one of the world's largest coral islands, also with landing grounds on its west coast like Aitutaki and Malden Island, but with high limestone cliffs and a large tableland. It has been known as 'the rock' since ancient times, but in the vernacular NIUEAN, the name may suggest 'behold the coconut.' Foreign boats on lookout for this landmark in the nineteenth century were primarily whalers. In 1887 it was still ruled by the native PATU-IKI (king) TUI-TOGA, also known as Ta-tagata. This old Polynesian kingdom was finally ceded to the British Empire in 1900. Today the autonomous territory Niue is known for its beauty and for its widely used internet domain .nu which is also its country code.

'The guano fields occupy most of THE ISLAND OF MALDEN, which is about fifteen miles in circumference. Here and there the coral island is grass-covered, and a few poor dwarfish trees are also found. Of course, we undertook walks and went round the atoll and "visited everything worth seeing" there. But I was never particularly amused by being ashore in the scorching sun. It was terribly hot the day we went up to the guano fields on the other end of the island.[103] We went in a kind of *draisine,* an inspection and transport trolley, which was pulled by natives.

We then became acquainted with the only "ladies" who live on the island, namely the wife and little daughter of a native foreman. With them we had breakfast. The young wife, who got married already at the age of twelve, looked terribly "pretty", in a way, extremely nicely equipped as she was in a fiery red domestic dress, richly embroidered with lace, and add to the image the dangling gold earrings, and fingers full of rings. It is said of her that she once saw a Melbourne fashion journal and became awfully fond of the then prevailing fashion, "the camel hump behind"—i.e., the *tournure,* the "bustle"—and that she gave her husband no peace until he sent to Melbourne for one such outfit to please her.'

33. Out Fishing with the Islanders

'We collected several curiosities at MALDEN ISLAND, such as beautiful, strange shells of several kinds, corals, bird's eggs, etc. However, what made our long stay there reasonably pleasant, was that we had so agreeable company. As already mentioned, there were two other Gefle ships there: The *Chili,* commanded by Captain Sågström, and the *Daphne,* with Captain Pettersson on command. These two names deserve to be remembered in my "journal of travel memories". They were really very kind, friendly, and nice old gentlemen, "uncles" to us.

---

[103] Ian Fleming's *Dr No* (1958) pictures a guano mine and loading wharf of this kind. Back on the ship the trying loading of guano went on with lighters and tubing in clouds of choking dust. Smoke belched out of the hatch and all covered their faces with rags, shawls, and balaclavas, looking like brigands. The point was to fill up the hold with the valuable substance, which paid the trip and the salaries.

On the *Chili*, we renewed our acquaintance with an old acquaintance from home, helmsman Rylander. Furthermore, we also liked a couple of the English on the island quite a bit: the *manager*, Mr Julien, and the general practitioner Mr Fox. We were in their company almost every day, and they tried as hard as they could to make it pleasant for us.

One of the last evenings we were out catching flying fish. I wish I had a real "painting pen", as it is called, to be able to give a vivid description of how this was done, and a graphic view of the remarkable scenery. If it were true to life and a well-wrought work, such a picture would be awesome, indeed. The backdrop and basic effect of the artwork would then be a sky sparkling of millions of stars, and the phosphorus-glittering sea. It was a real tropical evening that night (which in all its splendour still cannot stop the northerner from feeling homesick!).

But now to the fishing expedition proper, which would be the main subject for the paintwork on the canvas! With the one exception that we were going in the *Atlantic*'s ship's boat, instead of a South Sea canoe, it was all arranged in the usual way by the natives. The rowers were natives, and a couple of them held tar flares as torches and allure, illuminating in the dark, and others of them tended the bag nets. And it was very exciting to see how dexterous they were with their fishing-tackles, and we got a lot of fish. In actual fact the sea seemed to be completely alive and filled with fish, they incessantly appeared here and there, and the water glistened with fire and phosphorus in all their movements.

We rowed along the beach in shallow water, so we saw the coral reefs glistening on the bottom. We saw large shoals of sharks, dolphins, and springers, but fortunately these kept at a distance from the boat, while several others literally flew up into the boat; a flying fish even jumped right upon Mr Fox's nose.

A couple of times we have had the opportunity to send letters, once with the aforementioned Melbourne schooner, and once with the *Chili*, which sailed two weeks before us. What a long-distance travel for these envelopes in postal service! If we are lucky, and all goes well, then we might ourselves get home almost as fast as they do. Then they may serve as heralds of us.

We have been very diligent with our homework as usual. I continue to read English and German fairly attentively for a couple of hours a day, and intend to become really fluent in these languages.

Emy asks me if I still have plans to become a school teacher, which was my future prospects in the past. Now, for the moment, I read German in order to be able to reasonably get along in Hamburg, the nearest destination for our trip, although I must have forgotten to mention it here! Yes, obviously I do keep a travel diary, but completely forget to mention where we are going! But I have noted down that the ship will carry us homewards, and that is the most important thing!

A few days last week we were working ourselves almost to death with washing and ironing. At least I felt close to *done up* in the awful heat. However, although Malden Island is located only *four degrees south of the equator*, we never felt the heat there so dreadfully oppressive as in Australia with its hot land winds. A fresh, cool wind always blew, especially in the evenings and the nights, which were always admirably beautiful, in particular during this last week, when we had the most glorious moonlight.

We have been terribly tanned in the blazing sunshine, indeed; in skin colour we can easily compete with the indigenous *kanakas*.'[104]

34. Alteration of Course to South-East 10,000 km

—A quarter of the Earth's diameter at one go

The captain has decided not to go back to Australia a third time on this trip, but to head in the opposite direction instead to sell the fertilizer load elsewhere—preferably in Europe and by no means in South America where

---

[104] *Kanaka* (Polynesian) was a widespread Hawaiian word for 'human', *kanaka maoli*, or originally, 'free man;' *kane* means 'man.' *Kanak* in Melanesian, a people of New Caledonia. All but a racist or derogatory word at the time, it was by then used among late nineteenth-century mariners such as the Söderströms to refer to all people from the South Seas, even including the autochthonous population of Easter Island in the far east, and carrying a connotation of praise for the remarkable navigation skills of Polynesians—who had traversed the length and breadth of the Pacific in long canoes for thousands of years, and populated the Polynesian triangle (with the angles New Zealand – Easter Island – Hawaii), guided by the stars. The sides of that triangle are almost one-fifth of the Earth's circumference in length.

Chile now owned the world's largest deposits of guano and had secured a monopoly of the goods. The ATLANTIC was steering close south-eastwards to go by Cape Horn to the South Atlantic and then turn hard northwards.

'AT THE CAPE HORN—the horrible, dreaded CAPE HORN, which we must pass—I think we will get ample time to turn pale again! Yes, would that we were already over there, and on the other side of the globe, and our circumnavigation of the world were complete!

This is now my last trip with Dad. Next time I have to set out with another ship and with another captain on another journey! Oh, oh, oh! Oh, long-awaited time, when I get to see and embrace my own John again, can look into his faithful blue eyes, and hear him call me "my little darling," and all his other warm-hearted assurances and marks of love. But now by all means *stop this!* before the emotions overwhelm me and overflow all banks! Once getting into a chapter about my lovely and loveable sweetheart and his, in my eyes, amiability beyond measure, I would be able to produce volumes. Yes, "he" and only "he" is the beginning and the end of this story, this ballad or song of songs!

Emy once asked me if I did not intend to publish this large, thick book in print. (She imagines, of course, that it is terribly interesting, since she never has read anything in it!) Well, if that happened, it would certainly raise *furore!* Oh, what if someone unauthorized came across it and read it! How ridiculous and chatty it would not seem! Who would even understand it, all this jumble of thoughts and feelings! Who could really understand me? Personally, I understand myself perhaps the worst. I consider myself as a person without all the order and context. But for the future, I'll leave it to my John to keep track of me. I think he is fully competent for that; (may that task not seem difficult to him, I want to try to make it as easy as possible!) he has a great influence on me, I feel that. And he is so old and sensible, almost twice as old as me, 34 years old! Imagine that I get such an old and steadfast man! I'm really proud of that! And I myself am just a foolish child! Yes, if only I were childish in the right way! Childish and innocent! Oh, I wish I had all sorts of good qualities for John to value me so much more. I could not bear that he had reason to harbour a single bad thought about me. And yet I do not want the saying "love is blind" to be applied to him. No, may he see my faults and correct them! Oh yes, he has done so already! I received several admonitions, because he noticed that I can be capricious at times. I am infinitely grateful to him for that!

"And what title did you want to give your book in that case?" asked Emy. "The Memoirs of an Angel"? Unfortunately, that would be a pathetic parody! "The Autobiography of a Madcap" would perhaps be much more to the point, but almost too hurtful for my self-love to cope with! Well, quite simply: "A Girl's Memoir" would definitely fit in just fine. The word "memoir" must of course be included in the designation, it sounds so educated French and sensational, which a book about a girl's heart must be! ! ! Bosh! Gobbledygook! Poppycock! Daub! Babble! Drivel !!!'

35. Storm damage—With a Leak Towards Cape Horn

'THE 4TH OF JUNE, 1887 [SATURDAY]

We are both surprised and appalled! Dad has told us today what we have suspected and feared for a while, that the *Atlantic* has a severe leak, which after the violent storm of recent days seems to have increased significantly, so we have to look for the nearest port. It has been decided we will go for Valparaíso, South America, as a port of refuge. The distance there is still very long, and we will probably have several weeks more of travel.[105] In order not to worry us, our poor father tries to be as calm as possible, but we have perceived certain words and allusions between father and the helmsman, that danger is imminent and we incur the risk of not even getting there.

Now it is time to trust in God and Providence. Oh, I have had a hot battle with myself, and tried to persuade myself that if the terrible,

---

[105] 6,053.24 mi (9,741.74 km). The captain had calculated to go the well over the 6,814 nautical miles from Malden Island to Cape Horn in a smart bent course of over 12,000 km with ten knots in a little more than a month's time, but the wooden ship was too heavily weighed down with the mineral sacks, and the favourable winds too strong. They were close to turning over, and the creaking hull began to crack and they made water. As a comparison, in September 1884 the ship had gone between Söderhamn, Sweden, and Senegal (5,865 km = 3,161 nautical miles) in 30 days, which was considered comparably fast at the time, but now the bold captain aimed for the double speed, and failed.

feared thing were to happen, I would be prepared for it, ready to meet God—to die. There was a time when I longed to die. It was at the happy confirmation days. I then thought that nothing would be more fortunate than to die young, before the heart had become too attached to the earthly, and rooted in the mundane, and I know that my heart was then seriously attached to my Saviour. Nevertheless, I fear that this desire to die largely stemmed from an already ingrained fear, that in the future I would be more and more misled away from the One thing necessary, i.e., I did not trust that God had the power and will to preserve me when I was lost.

And how do I do now? A matter of conscience! It's awful that I have to admit to myself, that I shudder at the thought of death. Oh, to die now that I have so much to live for! And yet when the time comes, the time determined by God, what does it help to resist? Willingly or unwillingly, we must follow. It is extremely important to have it so set with God that we follow willingly.

It is true, as Tegnér[106] says, that "on the brink of the grave" all the cares and affairs of life will appear so small and insignificant. Oh, why, why, have I not considered it more seriously, that our whole life should be a preparation for death? Now when I know that once, sooner or later, it must happen! And now maybe it will occur very soon. God, my God, teaches me to find a friend in the "King of Terror."[107] To die in order to live with the Lord forever is not to die. Thanks to Jesus, death for His faithful is only a sleep. The awakening will be in heaven. Make this alive for me, O God, through your Holy Spirit.

> Keep me ready, O dear Jesus, to await
> Your sweet arrival, when you at last want to bring up
> From the Vale of Tears and Agony
> Your bride to the bright hall of heaven.[108]

And now, dear faithful Lord God, I leave myself and all mine and all of us here on board entirely under your mighty protection! You have

---

[106] Esaias Tegnér (1782–1846), Swedish national poet.
[107] *Förskräckelsens Konung*, the name of Death in *Job* 18.14 in the Swedish *Biblia* of King Carolus XII, ed. 1873, carried by the ship's library.
[108] From *Sionsharpan: andeliga sånger*, Oskarshamn 1882.

said: "Fear not, for I have redeemed you; I have called *you* by your name; You *are* Mine. When you pass through the waters, I *will be* with you; And through the rivers, they shall not overflow you." [Isaiah 42.1-2]

If we may understand this literally, that you graciously want to deliver us from this distress, then be to You eternal thanks and praise! Then we can also take this wonderful word as a confirmation that you have decided to carry us safely through the waters of the river of death, that its currents may not drown us, but that we may reach

the glorious shore,
which no storms reach.

THE 8TH OF JUNE, 1887 [WEDNESDAY]

A terribly stormy day! The ship is tossed about and wriggles, the wind is veering and the helm shifted. *The water pumps run all around the clock!*[109] We're terribly worried. Dad is also worried, we can see that. Now in the evening the storm increases really hurricane-like. But we believe that precisely because of its ferocity it cannot last long, and the barometer goes up a little!!!

I'm trying to escape from all this horror by dreaming. Of course, the thoughts fly to all the beloved ones far away, but thinking of them

---

[109] This momentarily meant *All hands to the pumps!* However, like most Swedish-Norwegian ships, the *Atlantic* was, in addition to ordinary manual pumps, equipped with *wind pumps* that freed the crew from pumping duties in hard wind, but these pumps did not work in calm [Leslie, *A Sea-Painter's Log* (1886), 52]. The increased weight of the ship, attributable to the instreaming water, reduced the net buoyancy, and moreover endangered the valuable cargo. In a situation like this, when the hull was leaking, *pumping* and *stopping* were the two measures. The first goal was to reach an *equilibrium*. As the boat fills with water, the rate of water entry slows down when the leak is under the waterline. Hence, a point can be reached at which the rate of water removal by pumping equals the rate of water ingress—then the ships stays afloat and does not sink. The next immediate step was plugging and seaming from the inside with oakum, canvas, tallow, pitch, wood, and other available material. The carpenter and many hands were busy in lower deck and hold and down to the keel. Essentially, this work was stuffing seals with packing tow, which had to be done regularly in a wooden vessel [Oertling, 1996]. When the outflow exceeded the inflow, then the captain sounded the *all clear!* Emergency port next (Valparaíso). Until then, '*the water pumps run all around the clock*', as the diary notes in the stormy weather.

makes me even more heartbroken. God forgive me, that sometimes I feel really defiant in my mind. Oh, I think it would be so cruel if we now had to say goodbye to life. Is God loving? God, you see my doubts, you see the hardness of my heart. Lord, subdue my proud mind, beat down the waves of trouble as you also have the power to calm down the waves of the sea. Oh, do it, save us, save, if that is your will! Lord, this is a crucial moment, I feel it. My soul wrestles with you; O Lord, become too strong for me! "I will not let You go unless You bless me!"[110] Give me your peace for Jesus' sake. Lord, I am yours forever, in life and death!

And my John, whom I now know is even more precious to me than ever before (*Oh, the feeling is as if the heart were torn from my chest at the thought of him, I cannot think of him*) I commend to You, O God. And our beloved old grandmother! Maybe my heart can find some rest and comfort if I ponder her last words to us: "if we are not allowed to meet again here on earth, we will know for sure: once in heaven!"

> There is my Saviour, there is my home,
> There I will see Him, live in His home,
> There I'll see next,
> The ones I loved the most
> There's me forever best,
> The blessed home.
>
> I'm a pilgrim here,
> going to my home
> A world full of troubles
> is not my home,
> Here on a stormy strand
> I am in the straits at times,
> But look, the good land
> Is beckoning me home!"[111]

---

[110] *Genesis* [1 Mos.] 32:26.
[111] "Jag är en pilgrim här." Quoted from the author Lina Sandell's Swedish rendering (1860) of the well-known hymn 'We are but Strangers here, Heaven is our Home' by the British psalmist Thomas Rawson Taylor (1807–1835), the son of Rev. Thomas Taylor. His hymns and poems were published posthumously in *Memoirs and Select*

36. Passage of the Horn at the 56ᵗʰ Parallel South

'THE 19TH OF SEPTEMBER, 1887

Times and circumstances have changed much since I made my last notes. Praise be to God, that the dark miserable time is over; now everything is bright and auspicious again. The danger that appeared so great and imminent has been turned away from us by the loving God; He has the power of succour in times of greatest hardship and distress, to relieve from menaces, and so He did with us. Just trust in the Lord!

Against all anticipations and suspicions, we were able to continue our journey around the Cape Horn! After the severe storm, which I described, we had excellent weather for a few days. When the gales had abated in the following mild and calm period, the leak was closely examined, and, to the utmost possible extent, attended to. To our relief and delight, our father and the helmsman then finally determined that the damage was not that serious and hazardous as they first had thought. After many deliberations thereafter, it was accordingly decided that—with God's help—we would venture to continue our journey, especially as the wind was favourable, and seemed to encourage this decision and the voyage.

Overall, we felt enormously happy and grateful, and for so many great reasons. When all is said and done, the journey has gone excellently well, and also remarkably fast, and now—joy to excess!—we are in the English Channel, and thus we are not very far from the destination. With a good wind we could be in the horribly long-awaited harbour already within a couple of days, but we are not that lucky, though; the last few days we have had a stubborn headwind, so that it goes so terribly slow—yes, far too slowly to our longing and impatient hearts.

> On 12 July 1887,
> *Cape Horn was passed.*[112]

Remains, 1836. This very hymn was written when he was at death's door, giving up the ghost from tuberculosis at 28.

[112] Particulars about the short anchorage at Valparaíso (for repair, victualling, and fresh water) are missing, as is also recorded time lag and delays due to the damage on sea, so the pace cannot be safely calculated, but it appears that the main speed probably did not exceed five knots, even if it were a virtually direct cruise of 10,164

Down there it was not as dissuasive and frightful as we had expected and feared. On the contrary, we were met with sunlight and beautiful weather during the passage. Consequently, we had unnecessarily been anxious and alarmed about that thing. Why do we procure imaginary worries for ourselves? Without figments of the imagination there is no fear.'

The long distance from MALDEN ISLAND *to* CAPE HORN *was covered in two months (fifty-nine days from 14 May–12 July 1887). It should be noted, that the captain deliberately avoided the narrow and rocky Magellan and Beagle Sounds for safety reasons, knowing the* WILLIWAW *winds, the number of* WRECKS *and* ROBBERIES *there, and went for the open-water* DRAKE PASSAGE.

'In any case, it was so wonderful to wend our way northwards from there, what a relief! Only then did we start to feel that we were really on our way home. It is indeed much more pleasant to be in one's own hemisphere. It is also more delightful and full of variety to sail in the Atlantic than in the Pacific Ocean with its immense expanse. Despite all its lush islands that we once in a while did have the pleasure of seeing a glimpse of, you are so desolate and alone there on the sea.[113] Here, on the other hand, we are surrounded almost daily by dozens of sailing ships. Very often we exchange signals with them.

It was a big "event" one day a few weeks ago when we recognized the Norwegian bark ship *Melanesia*, which had been in Melbourne at the

---

km between these two dates. This would indicate that the rate was similar to that of the reputably fast trip of thirty days between Sweden and Senegal in 1884.

[113] In his last appearance before the public before he died, the famous Italian writer Dante Alighieri addressed the ancient cosmological problem of the distribution of land and sea on earth, *Quaestio de aqua et terra* (1320). The factual basis of the problem was improved by JAMES COOK in the eighteenth century, who shoved that most land (in fact 68 per cent) is indeed found in the northern hemisphere as the ancient Greeks thought. In the nineteenth century it was evidenced that the Pacific Ocean occupies about a third of the world's surface, but the reason for this eccentric geological and oceanographic fact—which Dante tried to explain—the perpetual movements of the continents, was not known until the twentieth century.

same time as us. The weather was then very beautiful, although not very favourable for seafarers, namely *calm*, and the amiable Captain Klöcker was thereby able to put out a ship's boat and honour us with a visit.'

The large and newly constructed MELANESIA *was launched at Hansnes in 1884; built in Gårdalen, Strømsbo, Arendal, Norway. The life of this lavishly fitted out* BARQUE *was short. It was stranded and wrecked at* DELAGOA BAY *(off Maputo, Mozambique) in 1896. It was a timber merchantman like the* ATLANTIC.

'Of course, it would have been even more fun to meet someone who came from home who could tell us some news. Of course, a lot has taken place and come to pass in the world during the long time we have been away from it. For instance, what about all our loved ones back home? They have all been left in God's protection, so we should be easy in our minds concerning them. Have they perchance forgotten us? Oh no, not the best and closest friends. But have they even been thinking half as much of us, as we have of them? Has John longed for me a lot? Has he already left home and is out on the sea again? In that case, he must have been very worried and sad that he did not get the opportunity to learn that we had arrived? Questions that will all be answered in a few days! Until then: patience, patience!'

37. Portland ho!

'Portland, England, the 23rd of September, 1887.

So, now then we are in port, although not at our actual destination and port of arrival, Hamburg. The fact is that this morning we had to go in here to seek refuge from a stubbornly persistent easterly windstorm, which drove us back rather than forward.

Now they know at home by telegram that everything is fine and all is well with us, and we also intend to write to little grandmother to calm and delight her properly; poor little old woman, she has probably started to worry about us lately. Dad has also telegraphed to Hamburg to get letters and newspapers returned here, and it now depends on when there is a change of wind direction if we get them in time. In any case, we have learned a lot of news here, partly through English newspapers and partly through Swedish dailies, which we received from a Gothenburg steamer here. Oh, how I searched and "rummaged" them to possibly see the GURLI mentioned, but without results. Well, I find to my sorrow that she evidently set out long ago. Sure, it's annoying, but I have to content myself with the fact. Hopefully my darling has been really open-handed with the letters; yes, he must have written a whole package, it will be my compensation then.

SUNDAY AFTERNOON THE 25TH OF SEPTEMBER [1887]

We set out to sea again. We now have fresh benevolent breezes, which we set hopes on, and in a few days can go in to Hamburg.

We were not in time to catch any mail, which was a pretty big miscalculation, but we try to comfort ourselves as best we can and "take it philosophically" as Dad usually says, it can only serve to train our long-suffering patience a little more!

At long last we got to see England up close, albeit only cursorily. Emy and I have not been ashore here; it is a very small place, known for nothing but so-called "Portland's Cement."

Nearby are a couple of small towns, CASTLETOWN and WEYMOUTH, one of which has a fortress, which we see in the distance.[114] It is still green and lovely here, although we will soon be in October. Yes, this is the only glimpse of verdure we've seen this summer.

It has been really refreshing to finally be able to catch breath in a calm and safe harbour, after all the storm winds during our long journey. God be praised that we so happily escaped them all!

---

[114] *Isle of Portland* is a peninsula in Dorset on the British south coast (by the English Channel). Beside it is the immense *Portland Harbour*, the largest in the world, completed in 1872, which was their port of refuge in the headwinds, and a place of partial reloading and provisioning. From the pier they could see *Castletown* and *Weymouth* facing each other on opposite shores. Since the 1850s, the patented *Portland cement* had been in demand and in common use, exported to many countries, and in the 1880s even produced abroad under license. It was a valuable commodity. As a construction material and a constituent in concrete, it increasingly competed with the wood industry and the goods the *Atlantic* and other Nordic merchantmen usually carried.

IV. The Voyage in Context and Retrospect

1. Origins

THE OCEANS MAKE UP THE GREATER PART OF THE GLOBE, to the extent that the planet Earth may as well be renamed the Sea. In our world, life and water are biologically and historically interdependent. The climate crisis is all about water. The original water circulating in our blood vessels and passing lachrymal ducts, with mineral content reminiscent of its primeval oceanic source, may contribute to the attraction of coastal areas and seaside resorts. Above all, the Seven Seas have inspired freedom seekers and migrants for thousands of years, and made ports—like the ancient Carthage, Guangzhou, and Chittagong—the bases of cultural progress, world trade, and discovery.

Ancient Egyptians, Chinese, Indians, and Semitic Phoenicians travelled the oceans. Polynesians voyaged the Pacific guided by the stars. Mesopotamians journeyed the present-day Persian Gulf and the Gulf of Oman in the Arabian Sea and kept up trade relations with the flourishing Indus Valley in present-day Pakistan.

Seagoing civilizations, such as the mythic bronze age Atlantides of Plato's famous legend, appear to have connected the Azores in the Atlantic with the eastern Mediterranean, as similar contemporary structures and symbols all the way indicate a unified empire or an extensive cultural exchange—in a hitherto largely unknown past where civilizations rose and fell and succeeded one another. Likewise, ancient megalith structures, most of them found in coastal districts, indicate worldwide communications.

However, traces of wooden artefacts from prehistoric times are rare archaeological finds in the seabed, as are other organic materials handled by sailors—ropes, cloth, leather. We merely see the stones of the ancients. Nevertheless, pictures in rock carvings fire the imagination, as do ship tumuli, stone monuments, cuneiform writings, pictograms and pictures in desert papyri, and excavated underground tomb imagery.

Interpretation is the subject of contention, and knowledge is limited to available sources. The obsession with religion in the Middle Ages was a long-lasting illusion of posterity in view of the fact that most written sources were produced in cathedrals and monasteries.

Likewise, the ancient Egyptians' pessimism and obsession with the afterlife have been deduced and construed from sepulchral inscriptions and paintings, whereas later discovered contemporary prose stories and poems from the world of the living—from this side of the grave, so to speak—indicate a positive attitude to life.

Reading history backwards, the geologist and the palaeontologist peep through the holes they bore in the earth. Pondering on an excavated artefact, the historian tries to read meaning from the context of layers. An archaeologist of a future millennium sifts the dust of the twenty-first century—what will that unknown scientist know of our time? 'Look! There is a lonely object seen at times before: a mortar made of stone.'

Written sources are not mute, however. Thanks to the number of eminent historians, geographers and storytellers of Antiquity—from Homer and Hecataeus to Herodotus, Megasthenes, Diodorus Siculus, Strabo, Pomponius Mela and Ctesias—we know that the ancients were wandering explorers and itinerant voyagers, and that the pre-Columbian world was intersected by intercontinental sailing routes.

Naval historian Klas Helmerson, former head and organizer of the world-famous Vasa Museum in Stockholm, laid down an established fact in the following précis, pregnant with significance: 'For thousands of years, from old Egypt until the middle of the 19th century, sailing ships ruled the oceans. It was under full sail that Queen Hatshepsut's ship arrived at the Land of Punt in 1400 BC in search of precious goods, that Phoenicians and Greeks traded and battled in the Mediterranean Sea and that Vikings headed west and set foot on Iceland, Greenland and North America' [Helmerson, 2008].

2. Trade Routes

With the exception of the weather-driven, the prerequisite for all this progress and success on a global scale, was the regularity of air currents through the ages, in particular the perennial trade winds. In the 1600s, *blow trade* meant 'blowing steadily in one direction', which made people confuse the wind with the commerce that depended on it. Trade winds encircle the earth and were essential to navigation in the Great Age of Sail, from the middle of the sixteenth century to the late 1880s. The circular movements of winds and currents in the Atlantic were also vital to the triangular transatlantic slave trade, which made merchants in Bristol, London and other West European seaports rich. The whole idea of this sinister arrangement was based upon navigational conditions.

Winds tend to blow southward from the English Channel and grow toward West Africa, and then the trade winds turn right with a speed of some four to six knots, eight days of ten, mostly supported by the similar circular sea-currents heading in the same direction. That is why early South European explorers from Columbus on, as well as later merchantmen, ended up in the Caribbean, which determined the fate of that region and the American continents. A northbound return route with favourable winds clockwise completed the business circle, bringing sugar cane and colonial products along with silver and gold to the economic system of the empires in the 'civilized' Western world [Hallengren, 2019].

After the fifteenth century, this route had become routine, and all followed in the wake of the Italian-born Spanish explorer with simple mnemonic rules, and with nature's support: 'Head south until the butter melts, then turn right,' or; in the Roman tongue of the ancestors, *Ventis secundis, tene cursum*: 'the winds being favourable, hold on to the course'.

Hence, the use of large square sails in those days, a rigging for tailwinds. One of the hazards in the days of sailing ships has been incorporated into an English expression, though. A danger faced by crafts with square sails was abrupt shifts in the wind to come up from a sudden squall, blowing the sails back against the masts, putting the ship in a vulnerable state and a dangerous position, the ship then *taken aback*. Headwind meant a setback and standstill—or worse.

One may regard the centre of the Atlantic in the Northern hemisphere as a huge vortex of wind and water moving clockwise around the Tropic of Cancer. Once you get from the European mainland past Gibraltar, Madeira, and beyond the Canary Islands, via the Cape Verde Islands to the West Indies, the journey may go almost by itself if you have the wind and luck on your side. This is the eponymous Columbus route and the basis for the European colonization of the Americas and transatlantic commerce. That is why sailing ships for centuries gathered in large numbers in European seaports, waiting for favourable winds that lead into this mighty whirlwind dance, and then one could see hundreds of full-rigged ships set sail at the same time.

This is the southern route between Europe and America, and still remains the most popular among lone yachtsmen in that part of the world. For the Scandinavians, the distance from the North Sea down through the English Channel to North Africa was the most difficult and perilous passage to cover, with storms and pirates and the privateers of hostile lands on the way. Deaths were always expected, and ships sank or disappeared, often with all hands and without a trace. Captain Axel Söderström's elder brother died on the same route.

These dangers lasted until quite recent times, and in other quarters too, such as to those running the Australia tramp trade. Of eight hundred merchantmen in external trade that were built in Gefle between the 1750s and the 1890s, over three hundred foundered or suffered heavy losses [Nordenberg, 2014]. Captain James Cook's famous eighteenth-century explorations of the parts of Oceania that was later to be named Australia and New Zealand, inspired numerous later circumnavigators and settlers. He was the first to choose an eastward course round the world. In the next century, eastward circumnavigation became a commercial route for ships like the barque *Atlantic,* and due to the wind systems in the Southern hemisphere, voyages from Europe to Australia naturally became trips round the world [Werngren, 1989].

Even the journey of 1885–1887, as recorded in Maria Söderström's diary, completed a circumnavigation for this very reason, and in point of fact much more—but for business reasons: Sweden–Cape of Good Hope–Australia–North America–Australia–Line Islands in the

Pacific–Cape Horn–the United Kingdom–Germany–Sweden. The captain had done this before and would go through this repeatedly, including even more troubles at the Horn with storm damage and leakage. His ship was usually exporting timber, deal and other woodware, but also trading coke, fertilizers, or other goods, including grain homeward bound.

The Grain Trade Route was one of the names given to this track that the nineteenth-century navigators of merchantmen followed in shipping between Europe and Australia and back. Due to the trade winds and the westerly air current that flow around the Antarctica along *the roaring forties*—between 40° S and 50° S, where there are persistent westerly winds—it was most advantageous to go around the earth eastwards. There are accounts of captains of full-rigged ships that lay for months and cruised around the Cape Horn without being able to pass the notorious Horn westwards, and who eventually gave up and found it easier and faster to scud and run before the wind all the other way around the globe. William Bligh, the well-known captain of the ill-fated *HMS Bounty*, was one of the able navigators who had this hard-earned experience and dearly bought knowledge.

The opening of artificial waterways such as the Suez Canal in 1869, and subsequently the Panama Canal in modern times, was to change merchant fleet itineraries fundamentally. However, the old so-called Grain Route is still the most advantageous for such large crafts that cannot go through the Panama or Suez Canals. In a similar fashion, large and bulky sailing ships like the *Atlantic* kept to the old route while it was profitable, because they could not pass the Middle East canal without tow or motor propulsion, whereas small or middle-sized steamers and motor vessels chose the new way.

This double service of steam and sail went on until the cut-off point when the steamers became more cost-efficient, and demand for sailing merchantmen accordingly dropped. The last commercial voyage on the old trade route took place in 1949 when the heavy four-masted steel barque *Pamir* of Mariehamn completed the roundabout journey by passing the Cape Horn, homeward bound for Åland; the ship was finally lost in a cyclone off the Azores in 1957. That was the dénouement of the Age of Sail in the history of trade and shipping.

3. Prevailing Sailing Conditions

In 1885, when our journey begins, the British full-rigged clipper *Cutty Sark*, launched in 1869 and the fastest sailing ship of its time, was out on the seas, sailing from its London homeport to Australia and back, passing the southernmost part of the continent of Africa, as the *Atlantic* did on its way Down Under, and to Oceania.

That famous and historic sailing ship, constructed in Scotland, today immobile and stationary at Royal Museums Greenwich as a magnificent and now restored museum piece after considerable damage, was not a competitor of the *Atlantic* in its time, as the clipper was then on the wool trade route carrying lighter goods.

Its fate is a story of the times.

It had been outmatched by the new steamer merchants on the old Shanghai exchange, even more so since the Suez Canal was already open when this full-rigged vessel was built, a course which traditional sailing ships could not manage without engines. They had to head for the Cape of Good Hope on their journey to the Indian Ocean and the East China Sea, making the African roundabout as their forerunners had done for centuries. Steam ships went from the Mediterranean to the Red Sea and the Gulf of Aden in a great time-saving shortcut.

The formidable and luxurious high-speed vessel, which could make a good seventeen knots at most, was already antiquated when it left the stocks as a masterpiece of boat building. It was one of the last tea clippers, soon to be used for other purposes. The great Age of Sail was over; while slower, less expensive transports of goods over the oceans with ordinary ships such as the *Atlantic* still gave good value for the money until the twentieth century. Even today, that is also true of slow, plain worldwide freights by sea. Owing to the cost efficiency of the commercial tonnage measured in cost per ton/kilometres (*tkm*), shipping is the cheapest way of conveying cargos over the Earth, environmentally friendly, and as compatible with sustainable development as ever [Hallengren, 1982].

4. The Gefle Harbour

The *Atlantic's* homeport Gefle (spelt Gävle since the 1940s) in Gästrikland on the Swedish east coast is an old fishing village and harbour at the Baltic Sea in the Southern Gulf of Bothnia. The present-day city is some five hundred years old, but the site was home to Viking explorers and traders a thousand years ago—people skilled at building ships and who were sailing far and wide, using their knowledge of the stars to navigate the seas. A significant ancient monument and testimony still standing in this place is a runestone carved in the eleventh century and erected in memory of Egill, who 'travelled with Freygeirr', a Viking chieftain, and 'died in Tafeistaland' (Tavastia in Finland).

From the 1600s on, Gefle was an important Swedish shipping town. In the 1840s, Gefle had become the leading Swedish seaport, in the middle of the century the town of 10,000 inhabitants had some 120 ships in operation and over forty shipping agents. Major occupational groups were seamen, shipowners, carpenters, ropemakers, sailmakers, and so forth; the sea being a source of income to a large part of the population. During the latter half of the nineteenth century, it was the major maritime centre next to Gothenburg, with a formidable fleet of sailing ships in international trade. A great number of the vessels, all made of Swedish wood, were built on local shipyards, and certain among them were masterworks of contemporary craftsmanship. A reason for the prominent position of the seaport was the proximity to primary industry and supply of raw materials, such as forestry and mining, ironworks and sawmills; and significant export commodities were timber and wood products.

The Gefle ships were numerous and well-known in the heyday of its Northern harbour, with worldwide connections. The Gefle schooner *Ragnar* (built in 1874) and its crew and captain are minutely and unflatteringly described in the Nobel Laureate and former crewmember Aksel Sandemose's novel *A Fugitive Crosses his Tracks* (1933).

Captain Nemo, in *Twenty Thousand Leagues Under the Seas: A World Tour Underwater*, intimates that parts of his astounding construction, the submarine *Nautilus*, are manufactured in Sweden—then so famous for its naval engineers and designers [Verne, 1871]. As a result of the good reputation of the Gefle-built ships, there was also an attractive second-hand market and a tempting trade-in value for these crafts, which shipowners sometimes could not resist. The outcome was that some of their gems ended up in circumstances most alien and unwelcome to the naval architects and their former agents.

This happened to the glorious full-rigged ship *Daniel Elfstrand Pehrsson*, in 1857 constructed by naval engineer Daniel Elfstrand[1] for the business house Daniel Elfstrand & Co. It was the boast of company and builder. With its 61 metres from stem to stern it was elongated as its name and the longest ship ever built in Sweden. The exemplary craft set off on its circuit to Australia, India, and two years later was on the guano route to Peru. After storm damage in the vicinity of Cape Horn, however, it made water and sought emergency port in Valparaíso, Chile. To the reader of this book, the incident and the course are familiar. With the loss of value, and the company at the same time becoming insolvent, the ship was put on foreign auction in 1861—the eventful year of the Civil War in America where chattel slavery was at stake, and the Emancipation of the Serfs in Russia.

Thereafter the ship sailed with different owners and under different names, all unknown to the former homeport. With a serious leak it arrived at Port Chalmers in New Zealand in 1874 and at last finished up a wreck off the coast in Deborah Bay. It was then named *Rosalia*, but on inquiry it became known that it had for long served as a slave ship—transporting forced Chinese manpower to the South American guano mines—then denominated *Don Juan*. That ghastly hulk offshore is still known under this name to New Zealanders. There are indications that the ship also served in the transatlantic human trafficking with Spanish-speaking crew, but it is known that it operated the China–Peru line in 1868–1873, for some time also named *Elcira Subercaseaux*. Once it arrived at Callao, the principal seaport of Peru, with almost four hundred Chinese men. The number of rusty iron fetters and shackles found deep down in the wreck's hold, testifies to the lugubrious past [Henricson, 1990, 47; Nordenberg, 2014, 88].[2]

---

[1] Axel Söderström's godfather.
[2] To make the wretched thing worse, the driving force that built the Chinese ramshackle shanty towns on the Guano islands was the demand of the gold that the slaves produced under the whip and which was marketed at six British pounds per ton, which made merchantmen from Gefle and elsewhere flock to the Port of Callao, the Lobos de Tierra and Chincha Islands—and this in its turn meant steadily increased output at the source and harsher labour conditions for workers of various origin in indentured servitude and slavery. Captain Söderström and most other deep-sea sailors had been busy running in this tramp trade. Most seamen saw the villages of

Fortunately, many vehicles did not play such fishy hide-and-seek games in maritime history, shady undercover business kept in a darkness that is hard to see through. For very good reasons, many Swedish seamen believed that changing the name of a ship would bring down ruin upon it.

The merchantman barque named *Atlantic* 1876–1911, in this respect was rather typical, although long-lived and in other ways too quite out of the ordinary. In the 1880s, the *Atlantic* was Gefle's largest sailing ship, with the weather-beaten commander Axel Söderström persistent at the 4.5 feet steering wheel for almost twenty-four years, which was a record then and there. Moreover, his ship was the city's last deep-sea sailer. When it was wrecked in a November storm in the North Sea in 1911, this marked the grand finale of an era, the definite end of the superannuated Age of Sail, replaced by the Age of Steam long before.

5. The Master Mariner

The captain of the *Atlantic* was an old salt of the past, known in contemporary sayings and literature as the Skipper King (Figure 18). He had spent all his years on the sea, had many crews, and was known in foreign harbours and by most captains and mates on the trade routes, foreign as well as domestic. He was seen decade after decade, returning over and over again.

In Sweden and Norway, which formed a union until 1905, and the union flag of which the *Atlantic* was flying, the nickname was supported by the striking likeness of this grave and awe-inspiring man to King Oscar II, who was bearded and balding similarly, and of the same height.

It was reported by his daughter Emmy that on a walk in the park Royal Djurgården in the capital, the old hunting grounds of the royal

---

tumbledown wooden hovels of the 1870s, but they had seen misery elsewhere and did not raise an eyebrow [Henricson, 1990, 43; Garberg, 1962, 58ff.]. In these times of poor crops, urbanization and overpopulation, the value of a shipload of this fertilizer could command a price comparable to the value of the old tubs that transported it. Greed and survival are instincts, and opportunity makes the exploiter.

Fig. 18. *The Skipper King. Captain Johan Axel Söderström in his prime.* Photograph owned by the author.

family, some unknown people he met with raised their hats in courteous greeting, believing it was the king on his daily regular. A man of peremptory authority and imposing stature, used to ruling: that was the image. Like Captain Ahab or the Sea-Wolf? Or was he copying the royalty, in the same way as his daughter Emmy adopted the hairstyle of princess Victoria of Baden, a descendant of Gustav Vasa who had arrived in Sweden in the early 1880s to become the queen? Was this Stockholm visit connected with a silver medal that the captain received for his service and sea measurements? The sources won't tell.

6. Loadstars

The captain had done brilliantly at school and college, and already in his early youth he had taken special interest in marine charts, star charts, and astronomical tables. To him, the seabed and the firmament were connected. Along with an ardent awareness of basic data of buoyancy, bearings, depths and fairways—the position in the universe—came meteorological studies, and all the more observations. Public weather forecasts for sea areas did not exist, nor any other daily news reports—they arrived backdated, if ever. To keep an eye on the sky was inevitable, and to listen to other mariners coming by and from afar. Above all learning physics: the movements of currents, winds, and stars. Even more so when the equator is traversed and the antipodal world opens, and all becomes reversed.

Seen from above, from the Polar Star due north, which is aligned with the inclined axis of our Earth, the globe rotates anticlockwise. Viewed from the Moon, it moves continuously from left to right, that is, goes eastwards, incessantly, as long as it exists. Seen from Down Under, the Southern sky and the Southern Cross, it is the other way round, the exact opposite. This rotation affects the currents of water, the air, and the weather systems globally, creating trade winds, vortices and eddies, high pressures and depressions, and cyclones and anticyclones. Relative to the stars, the Earth rotates around its axis with a velocity of about 1674 kilometres per hour at the equator, whereas at all other latitudes the speed is lower due to the gradually shorter circumference there. At Stockholm's latitude it is but 792 km/h (220 m/s). On the geographical poles, the perimeter then verges to zero (limit→0).

This difference in pace north and south has enormous consequences for world weather, as air masses with different speeds and temperatures perpetually confront and collide as the world spins. It causes the Coriolis effect, first described by Gaspard Gustave Coriolis (1792–1843), which makes the winds blow anticlockwise in the northern hemisphere and clockwise in the southern part of the globe during depressions. Accordingly, the air around areas of high pressure rotates clockwise in the northern hemisphere and anticlockwise on the southern. These systems of cyclones and anticyclones in the atmosphere have their counterparts and parallels in the oceans in streams and currents, in whirls and swirls on different scales, corresponding to whirlwinds on land. Everything is in a state of flow and movement.

Our captain along with mates and crew would agree with the proverb *Navigare necesse est*, 'it is important to sail', adding that it is intricate and perilous but would by no means agree with the Roman commander Pompey the Great's exhortation to the seamen, who hesitated to set sail in a storm to freight essential corn from Africa to Rome: *vivere non est necesse*, 'it is not important to live'. They made their living from carrying shiploads of heavy goods across the oceans through maelstroms, shoals and sunken rocks, fogs, overturning wind gusts and cloudbursts—a profession where it was difficult enough to save one's skin, and where fatalities were legion. Anemometer, measuring windspeed, and gauging speed of the ship by a log in tow, a float attached to a line on reel, were crucial calculation quantities of transport in an unendingly moving system of air and water.

The seamen had heard of the challenges of ancient predecessors such as Sindbad, Sinuhe, Ulysses, and many beyond, and knew the pre-Christian words of the *Qōhelet* (Ecclesiastes):

> The wind blows to the south
>   and turns to the north;
> round and round it goes,
>   ever returning on its course…

And the sun rises and the sun sets, and hurries back to where it rises. At sunset, the wind changes direction after an amazing moment of absolute calm when everything is still, and the surface is smooth as a mirror. Then silently grows the wind that blows all the bad air out of the land at night, followed by the wind that blows the sweet air in from the sea in the morning. Of all nature's rhythms, malevolent or benevolent, the spectacular calm at coastline sunsets was a revelation and relief to a seaman.

7. Astronomy and Meteorology

With favourable tailwind, they could once in a while go ten knots headway in daytime, or that many nautical miles per hour. If you depart on a sailing trip like our Söderström party, going well westwards with the east wind getting up, you in reality move eastwards with a tremendous speed, relative to our neighbours in space—the Sun, the Moon, and the planet system as a whole. So, taking this into account, you should add to that the velocity relative to our Sun: The Earth's orbital speed around the Sun being about 30 km/s (=108,000 km/h, ≈ 70,000 mph).

And, furthermore, the Sun's orbital speed around the galaxy: ≈ 200 km/s (720,000 km/h, 450,000 mph) where we are fellow travellers following along at the same fantastic orbital speed, which takes about

225 million years to carry out. This is our Galactic Year. Since the Sun and the Earth first formed, some twenty galactic years have passed; we have been around the galaxy twenty times, and there are more to follow!

On top of this, our galaxy, the entire Milky Way spins like an enormous wheel going at an incredible rate through the eons and, in its turn, in revolution in a much bigger system of galaxy clusters, guests from a long way away, with a long distance to go, which in turn are travelling widely in different pace, no one knows whither, or how these momentums should be construed. Maybe it is all a zero-sum game, the vectors cancelling each other out?

Leaving his books, tables, calculations, equations, speculations, and the schedule on the drawing-table in his stateroom, the navigator of the *Atlantic* turned to observation and steering by perception combined with intuition, gained from experience. The forces of nature were tangible. As commander and foreman, there were also substantial and corporeal conditions to engage on board, earthy things, and these things were always on his mind while he watched day and night, alone at the wheel in the wind.

Now, in the 1880s, Captain Söderström observed that the weather was rougher and the winds more unpredictable than during the preceding decades. The equinoctial gales were worse, and the frosty nights increasingly long and terrible at sea. It had become colder, perceptible even in cabin, bunk, berth, stateroom and saloon. However, the galley remained a source of heat, warmth, talk, and occasional cooking. There was an old-fashioned metal bin filled with sand and charcoal in which they sometimes lit a fire for frying or grilling al fresco, the cook's special treat with fish and chicken. On these rare festivities at approaching skerries and arriving in harbours, they moved to sea rhythms on spangled dance floors and embraced strangers and laughed into the long nights. Later, alone and lonesome, one of the young women thought, 'so disembodied we have become'. Shore in sight, *land ho!* Solitary in the middle of nowhere, 'may the wind and the pounding of waves bring us back to the iambic of our heartbeat and the swoosh of bird wings in our dreams'.

There was a firewood stove of cast iron on board, in early years a small tile kiln, soon lost, equipped with pots, pans, casserole and ceramic ware, a huge coffee pot and copper pan as chief items, furnished with a slim funnel of plate, but usually there was no fresh food. A lifeline was the food chests with dried salted beef and pickled herring, casks with dark small beer or mead, huge barrels of fresh water. But they caught fish also, some coming on their own by air as flying fish in the southern seas, others considerably more deep-going like the unsightly but abundant catfish in the northern sea. They hunted large migrant and ocean-going birds with nets, arrows, hooks, and deceit. The second mate and the cook always tried to get pigs or hens in the ports; occasionally a cow, which would be kept alive on board as long as possible. Sometimes there was such an awful row below deck that there would be the devil to pay as the sea rolled above and below.

The pattering rains that the hurricanes carried in tow were peculiar and did not add to the water supply in the swaying storehouse. The violent storm winds and the tropical cyclones knocked people down and battered things into pieces, and this they did equally to the air itself. It smashed molecules asunder, sorting out lighter and primordial isotopes as Oxygen-16 of stellar origin, making precipitation so different to be measured in ice and stalagmites thousands of years later. By the same token, the transformation by lightning of the normal two-atom oxygen into three-atom ozone was another outcome of the tempests, so severe and palpable to human beings alone on the open sea with senses on tenterhook and sensitive as seismographs, exposed to the weather gods.

8. Sea Rhythms

Normally, the sea rolls in its regular harmonic 100-meter cycle of waves, charged with energy by winds and occasional geological activity. Not moved by streams and currents, the water stays on the spot in elliptic circular movements, being over 780 times heavier than air and hard to move, rollers are large ripples on the surface, like sound is transported as waves through a medium without displacing it.

The heavy mass of water constantly puts hulls to strength tests due to the high density and linear momentum of the fluid—the ship's body challenged by wave pressures of six tons per square metre in stable weather on the deep seas, twice as much in rough sea. Rocking and tossing on top of the billowing waves in a slow and even up-and-down rhythm, and not breaking them or pitching, is the navigator's code of survival. The art of riding as a float in this big dipper set the limits of boat lengths in the age of sail, which should be considerably shorter than wave lengths (optimum ≈ 0.5 λ), and made it possible for single-handed yachtsmen, like the pioneer Joshua Slocum in 1895, to cross the oceans with smaller vessels.

There was another aspect to take in consideration too. In the middle of the 1880s, the atmosphere, oceans, climate and farming were affected by one of the worst natural disasters in historical time, the eruption of Krakatoa in Indonesia in August 1883, which cooled the entire planet and caused record rains worldwide, and after which sunsets were so red that people called the fire departments on the other side of the globe. For years it was reported that the moon appeared to be blue, and sometimes green, due to optical phenomena in the sulphuric haze. Bad crops and misery are invariably the longstanding outcomes of such catastrophes along with palpable and sometimes fatal weather changes.

Tales from the high seas about its strange forces abound among seamen, and many stunning tall stories from the tall ships have been verified. In his captivating book 'The Monster Wave' (*Monstervågen*, 2020), the writer Lars Berge tells about how the Swedish sailor J.W. Granström in a violent storm was washed overboard from the full-rigged Norwegian sailing ship *Alcides* on its way round the world, and how a big wave then washed him up on deck again! Investigating the truth behind this tale, the author found that similar miracles in reality had been reported many times from the seas.

This reflects a fact of much larger dimension: the number of mariners washed overboard in storms, of which a handful fortunately turned up again—as in the myth about how Jonah was swallowed by a 'big fish' (Hebr. *dag gadowl*), but saved after three days and three nights in the underworld, according to the ancient *Book of Jonah*, which in figures of speech with universal significance pictures the agony of the seaman on the brink of the abysmal sea grave:

> Then Jonah prayed unto Jehovah his God out of the fish's belly... Out of the belly of Sheol [the underworld] cried I, *And* thou heardest my voice. For thou didst cast me into the depth, in the heart of

the seas, And the flood was round about me; All thy waves and thy billows passed over me. [...] And I said, I am cast out from before thine eyes [...] The waters compassed me about, even to the soul; The deep was round about me; The weeds were wrapped about my head. I went down to the bottoms of the mountains; The earth with its bars *closed* upon me forever [...] And Jehovah spake unto the fish, and it vomited out Jonah upon the dry land.

In his overview, Berge gives a penetrating survey of the life of boys and men who lived and died on the seas. Some found their sweetheart in a seaport. Others drowned in alcohol. Many found God, but more were lost at sea. All went to female fortune-tellers, though, at home and abroad.

9. Life on the Sea

We travel backwards in time to the unwritten past in our memories, and onwards to the unknowable, undetermined future in our imagination. To the annalist and chronicler, time travel is a calling. History is like mathematics in so far that it is not a science, both being among the most extreme of human calculations in the sense that they are solely based on reasoning and imaginary worlds, not on perceptible reality. Keep in mind that the past does not exist anymore, it's not there—unless seen at an astronomical distance from another part of the universe. Not by us, then; whereas the loadstars of our navigators show what they looked like hundreds of years before. If the North Star exploded three hundred years ago, we would not know; it is where it is, as if eternally, yet is an illusion as any dream. Stargazing is looking into the past while the present is unfamiliar and unknown.

A nineteenth-century navigator knew the vault of heaven, the starry sky, and the zodiac north and south, better than other people. In daytime, at noon exactly, determined by the ship's clock, old mates and masters on the sailing ships measured the altitude of the Sun with the sextant to establish the exact position on earth—first the latitude, given instantly, by the date. To calculate the longitude was always more difficult, but by comparing local time on board with a chronometer's zero-meridian time, the east–west position could be looked up in a table. At times, this complicated manoeuvre had to be brought about at night, using a bright star as the guide, a measure involving astronomical tables and mathematical equations.

To choose the right course, then, was not just a matter of compass direction, because the chart showed a projected flatland with obstacles on the way, whereas the earth is round with no strait lines; this is why considerable time at the drawing-table was called for. Like the land surveyor, the seaman was a geometer well up in the trigonometry of triangles, and knew his Euclid and Pythagoras's theorem as well as how to make use of the observed angle between celestial bodies.

The world of the past is full of life and stories that are brought to us and kept alive in the words of songs and poems. At times they may contain the sharpened perception of an oral stage of bygone days with a more direct and genuine perception and taste of reality, with few intermediate agencies or the bucketing flow and streaming noise of modern media.

> So very much having passed before our eyes
> That our eyes in the end saw nothing

—the Laureate poet Giorgos Seferis observed. That was not yet the case with these old-fashioned mariners, who watched the wonders of the world with awe, apprehension, and fear. They were aware of any change or touch, however slight. Just like a minimal tremor in the cobweb is discerned by its custodian as either a puff of air or an opportune guest, and low-frequency juddering of looming elemental forces put birds and mammals to flight in the lull before the disaster.

A difference between ancient mariners and the 'primitive' peoples they encountered, was that they did not adore nature, whereas both parties were dependent on it for better or for worse, and had their share of its capricious benevolence and malevolence.

Blind faith, fatalism, and superstition were common to seamen and islanders, even more so when lives were ceaselessly lost at sea. At fishing hamlets, docklands, in archipelagos, and among littoral populations in general, religious worship thrived synchronously with the rising number of widows—sons and husbands drowned or disappearing without a trace—only flotsam, jetsam, and driftwood washed ashore long afterwards telling the stories of the dear ones. It is in this gloomy maritime context that the religious ruminations in the travel diary should be seen.

On the high seas, contrariwise, women on board were of old considered calamitous and ominous. They were thought to bring misfortune to ship and crew—while colourful bare-chested mermaids as figureheads at the bow of vessels were meant to calm the underwater merman and win his good will.

Consequently, partner relationships, mating, and family were all exclusively shore-based matters. A ship had *feminine gender* and was referred to as *she* and *her*, irrespective of its honourable baptismal name. The grand partnership on the sea was that between the masculine crew and its vessel of the opposite sex. That was *us*. On the opposing side they had Nature, friend and foe, with all its forces to bridle.

In the nineteenth century, women were considered out of place or inappropriate in many occupational branches. Female crews existed though, and captains at times brought wives and daughters on board as guests, many of whom volunteered as helpers or assistants in different ways. However, all in all they were still considered somewhat misplaced, *malplacé*, and were outsiders off the record, usually not signed or enrolled officially.

Later, when steamers in the merchant fleet introduced female cooks, these created a sensation and were at first proudly presented and loudly greeted in harbours as 'the Swedish Flag'. These flagship trophy women then replaced the old curvy figureheads. In older times, women in the crew were banned and would remain untouchables.

There were also certain things that you were not supposed to say and do aboard, and which would bring bad luck if you did. In addition, there were *non-sailing* days such as Thursdays (the day of Thor, the god of thunder) and Fridays (when Jesus was crucified); this is why the *Atlantic* started on a Saturday, which Shabbat observers would not do! Furthermore, travellers in the Pacific Ocean picked up Polynesian words that entirely answered to universal usage and custom and became colloquial, such as *taboo* and *tattoo* and even the supernatural power of *mana*, which remained well-known and inalienable elements of the sailor's world.

It is in this full context that we should regard the appearance of a couple of teenage girls on a sailing ship—travelling for years in a Man's

World—an otherwise unisexual environment with all its prejudices, ideas, and desires!

The standing and rank of such young girls onboard were that of cabin boys, cook boys, deck men, and factotums, and served as the officers' girl 'Fridays'. What is more, they were well-protected decoys and pranksters, modest Saturday night entertainers and regularly grave Sunday psalmists, always dressed in black or grey, and wrapped up well.

10. Authority and Management

Their presence—and these particular circumstances, customs, concepts, and conditions—added to the many reasons why discipline on board was mandatory, the captain omnipotent, and insubordination punished.

Obedience was taken for granted, but the master of the *Atlantic* on occasion locked his door for the care of private arms, and, as called for by the damp sea air rather than by usage, cleaning and oiling his new weapon, a Francotte à Liège M/1871, a heavy 11 mm revolver manufactured by Husqvarna, Sweden, known for its wood stoves of iron, of which one was on board, and later its present-day bicycles (invented in 1885). It was a modern handgun, this model invented and made in his homeland, loaded with central-fire cartridges with cases of brass and 13.2 gram lead bullets.

There is no evidence to prove that he ever fired, but one or two of his later mates and armour-bearers did, like blazes, and then smoke of black powder and sulphurous and nitrous fumes settled upon the ship to quell rebellion on board.[3]

---

[3] As firearms were as common with officers as revolts and attacks on the long trips, there were revolver shots and bullet holes on the quiet and beautiful sailing vessels, the swans of the blithe breezes. Moreover, the contemporary Wild West gunmen had competitors on the waters. Captain Fredrik Miltopaeus, who preceded Anders Clase and Axel Söderström as the skipper of the *Gustaf Wasa*, is known to have fired at his men in the rig, acting as a lion tamer, and for beating his mate. At home, he and his wife were known as pious and respectable church-going people; on the seas he was the notorious sea devil. Under the command of Anders Clase and Axel Söderström, the voyages of the *Gustaf Wasa* were purportedly more peaceful.

A very telling story illustrating the respect and obedience that was instilled by the master's acknowledged authority as absolute ruler can be found among the sisters' seafaring descendants and was handed down to the present writer along with some other observations, recollections, and witness reports [Herrström, 1973]. It runs as follows:

'The captain stood dead at the helm, with his eyes wide open staring straight in front of him, motionless and immobile, not turning the steering wheel anymore, while the ship was at full speed ahead approaching land and going straight to the shore.

He did not give any orders, did not react, and no one dared to do anything or act, and the vessel ran aground with a terrible roar and was wrecked.

The master had suffered a cerebral haemorrhage, the doom of many of the elderly mariners, and he remained erect unconscious and paralyzed, impaired, with fixed but vacant look, while the craft under his command crashed and went down.'

This was the fate of naval commander Torsten Bure Herrström (1900–1954), working for the big shipping company Nordstjernan, who died at his post outside Örnsköldsvik in the Gulf of Bothnia aboard the disabled and damaged M/S *Colombia* of 7060 tons deadweight, which shared his hard fate on 24 November 1954.

The crash was reported in the largest Swedish dailies next day, the *Dagens Nyheter* and *Svenska Dagbladet*. Attempts to bring the ship and the man to life were a dead failure in both cases. At the final break-up of the wreck of the M/S *Colombia* in 1955, the ship burst into flames and the fire could be seen for miles and miles along the coast.

The adventure of the dead sea captain is food for thought and a suitable subject of attention here. The *Colombia* was a diesel-powered motor ship of iron, travelling in a later era, yet the captain's authority remained the same as that on a wooden sailing craft in the century past.

11. Embarking the *Atlantic*

Captain Axel Söderström was known to have a calm and unobtrusive, even careless way of management, and his leadership was built by experience upon a strict entrusting of responsibility and job sharing among the crew. This was coupled with a minute distribution of tasks to everyone for all parts of the trip, made clear from the start, including passages through atmospheric depressions, erratic doldrums, and dead calms; all had their specified duties.

His daughter Emmy Söderström remembered how they all were set to work and kept busy at prolonged, dreary windless periods and standstill—in various repair work on hull, masts, sails, cordage, and in interior spaces; caulking, cleaning, grinding, tarring, paying, lubricating, painting, varnishing, polishing, washing, sweeping, and so forth, even if parts of this had been done recently and might appear superfluous or redundant to newcomers. But then again there was always work to do on a wooden windswept saltwater ship in the deep sea with heavily moving cargos downstairs, such as logs, board, deal, mineral salts, coal, and bar iron.

The master's motive, in keeping all busy as usual or more, was to keep the crew as calm as the sea.

Business as usual also meant that he had paperwork to do in peace and quiet. The captain was a salesman whose call was to keep his merchantman afloat by creating profits for the owners, which paid salaries to the crew and maintained the ship in return. The boss was a middleman in economical transactions without end, a link in a line of sellers and purchasers at home and abroad. By some mates he was remembered as mostly low-voiced and reclusive, one who preferred to be alone at the helm in all weathers, trusting the wind sensitivity of his cheeks rather than weathercock, windsock or gauge, to find the exact direction of the headwind, and then determine the course, and even the setting, hoisting or taking in sail, based upon that personal perception. Then he took his bearings to see how the land lay without sounding out opinion.

This atmosphere, which he appears to have created for himself, was based upon the rigid rules and discipline on board, set from the beginning and meticulously monitored. It takes time to attain such order, and he had come the hard way, maintaining a tradition of forebears as apprentice, learning from their accomplishments and experiences as well as from their failures.

He was the last in a long chain of masters. His grandfather Jonas Söderström (1774–1810) was Admiralty Lieutenant and captain in the merchant navy, dead at 36. His father Jonas Reinhold Söderström (1801–1862) was the master of a suite of notable ships, including the schooner *Sirén*, the barque *Minerva*, the frigate *Blixten*, the schooner *Hoppet*—meaning '*The Hope*', the most common name of ships from his homeport!⁴—and finally the barque *Christina*.

Fig. 19. *Captain Axel Söderström portrayed by Christophe Delphin Comergnac in Rochefort harbour, France, in 1882. Private possession. The Hallengren Archives.*

A list that may be impressive, or instil apprehension, depending on perspective. In fact, Axel Söderström's father Reinhold was accident-prone, and periods of his later career ill-fated. On one occasion, almost all the crew jumps his ship in North America. On another occasion, a tropical disease strikes his men in South America. In

---

⁴ We are told that the sailor Eric Ericsson, a confident man, in 1828 fell overboard from the Gefle frigate *The Hope* in the *Atlantic* and was considered lost; yet was picked up by a French brig three hours later, safe and sound, after having escaped the sharks and learnt to swim as a result of it—an exceptional skill among seamen [Henricson, 1990, 111].

IV. The Voyage in Context and Retrospect   **185**

harbours far abroad, he has had to hire complete strangers of various origin and worth. All this is evident from the preserved enrolment ledgers in the Gefle mercantile marine office, now in the Swedish National Archives.

However, to an old captain that was nothing. His son would, by the compelling force of heritage, continue on his course, but in early years revolted against the idea. He nevertheless picked up two things: to stick to the same ship and to employ well-known friends and relatives.

As an ambitious student at Gefle secondary grammar school up to 1856, he aimed for medical training to become a qualified doctor, a dream that did not materialize but left its lifelong mark on his care and nursing of crews. Among the boys in his class was Per Engström, who was later to become his shipowner, and he followed his father on the sea, compulsorily. In the summer of 1856, he appeared as cabin guard at the *Christina*, but when his brother Carl Gustaf perished at 27 in the North Sea on 9 October, he was determined not to follow course.[5] However, when personal items of the deceased long afterwards were retrieved and returned to the inconsolable mother Brita Maria Hammare, and the parents in affection conveyed the recovered overcoat of the deceased to the 15-year-old kid, the latter flied into temper and shouldered his fate.

His own course was now determined. The College of Navigation he passed at 19, scoring number three among the foremost of 205 students. As trainee and apprentice in his student days, he served as constable at the ship *Gustaf Wasa* under the command of his future brother-in-law Anders Leonard Clase; then he was deckhand (*jungman*) at the barque *Daniel*; and, finally, was signed as second mate of the ship *Andrea*, later on stranded. A successful student who became master mariner before his twenty-fourth birthday!

He was appointed captain of the full-rigged ship *Gustaf Wasa* from 24 August 1865, the vessel that the next autumn was to carry his 18-year-old wife Amalia Mathilda, and on which at least one of his

---

[5] The journal *Norrlands-Posten* of 6 November 1856 carried the late and decent news that the seaman Carl Gustaf Söderström 'quietly passed away aboard the brig *Juno* during a voyage from Gefle to Brazil on 9 October at 9 a.m. at the age of 27 years 8 months and 11 days, mourned and missed by spouse, parents, siblings and friends'. The demise of his father was announced in the same journal on 20 November 1862.

daughters were conceived. What joy! A couple radiant with happiness, sailing before the wind!

Fortune was to turn, however, when the all the more absent father took on the full-rigged *Thor* in the 1870s—followed by the untimely death of his wife in 1873, which made his children charity girls; and the foundering of his ship the year afterwards at the Azores (Figure 20), where the crew and finally the captain took to the lifeboats in a storm, and narrowly escaped drowning—all at the last moment saved by the passenger steamer *Jacob Setterwall* from Stockholm, incidentally on passage.[6]

Then the period of probation was over, and the master was biding his time at the schooner *Nordstjernan*, 'the Polar Star', for the next challenge, which was to be his mission in life. In his prime at 34, he

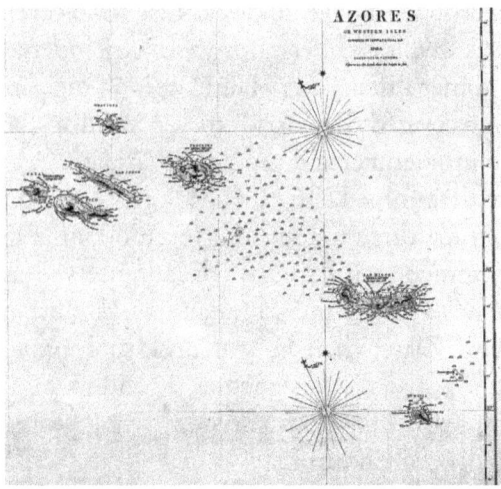

Fig. 20. *Nautical Chart. The Azores. Detail. Perilous waters feared by the captain.* Copy in private possession.

---

[6] The *Thor* was built by Pehr Elfstrand in Gefle for the shipowners Engström & Luth in 1855. Their barque *Astrea*, built in 1852, sank in 1854 outside Brazil, but the crew was saved, including the cook Eric Andersson Ljung, later captain Young, who became the maternal grandfather of Eric Linklater in 1899, and without whose rescue from the wreck there would not have been such titles as *Pirates in the Deep Green Sea* and *The Wind on the Moon*, besides numerous other changes in subsequent historical events, as would have been the case if the company of the *Thor* had been lost with the ship...

is employed as captain of a magnificent barque under construction at Brodin's Wharf in Gefle in 1876.

For once, that was a fortunate year in ship building. The number of Swedish ships built in 1876 and later lost, was none. Naval architects excelled themselves. Of ships built in 1885, there were only one lost after having been discarded: the HM Corvette *Freja*, which sunk in 1946 after the Second World War and became a wreck at the bottom of the sea.

12. Biography of the *Atlantic*—The Early Years

The *Atlantic* is a kind of barque, a smooth-running downwind sailing ship with three masts of which two, the fore- and mainmasts, are square rigged, while the sternmost or aftmost mast (the mizzen) has no yards but is gaff-rigged fore and aft. In other respects, the rigging is largely similar to that of a frigate or a full-rigged ship on which all masts have square sails. There were barques with more masts than the *Atlantic*, but the different rigging in the rear is common to all and is the ingenious point and the beauty of it.

The advantage of these rigs is that they need smaller crews than a comparable full-rigged boat or brig-rigged vessel, as fewer of the labour-intensive square sails are used. Another advantage is that a barque can outperform a schooner or barkentine, and is both easier to handle and better at going to windward than a full-rigged ship. While a full-rigged ship is the best runner available, and while fore-and-aft rigged vessels are the best at going to windward, the barque is often a good compromise and combines elements of the two. The efficiency of the *Atlantic* in various weather conditions and winds could also be improved by extra gaff sails, trysails, and staysails—as can be seen from the oil painting.

The *Atlantic* is built to be a fast freighter vessel in international trade and shipping, yet it is not really a *clipper*—from *clip*, 'speed' as in the phrase 'going at quite a clip'—but it is cost-effective. 'Clipper' does not refer to a specific sail plan as 'barque' does but could basically be schooners, brigs, or brigantines, most of them of smaller size, slimmer and less deep-draught, carrying a limited bulk freight, built in the mid-1800s specifically for fast deliveries of tea and other light goods from China. Famous is the Great Tea Race of clippers in 1866. Some of them could make sixteen knots on the average, playing in another division than the *Atlantic* and serving a completely different market.

The barque was in fact the most common kind of deep-water sailing craft in the nineteenth century, and this very type of vessel, with all its varieties as well as its name, was famed and familiar in earlier times. James Cook sailed the HM *Bark Endeavour*. The word 'barque' entered the English language via the French, which in turn came from the Italian and Spanish (and late Latin) *barca*, 'boat'. William Shakespeare uses the word 'bark' in the latter sense in his sublime 'Sonnet No. 116', published in 1609, when comparing the true love of man to the loadstar of any sailing vessel on the seas:

> Love is not love
> Which alters when it alteration finds,
> Or bends with the remover to remove.
> O no! it is an ever-fixed mark
> That looks on tempests and is never shaken;
> It is the star to every wand'ring bark.

Although rain after rain passed by on 10 June 1876, many people had gathered in the Gefle Harbour to see the launching of the new ship *Atlantic*, which was fully equipped and set sail a month later, given the ship list number 2787.

Main shipping office and owner: Engström & Luth. Length 57.18 metre, breadth (beam) 10.26 m, depth 6.65 m or 22.20 feet. 1,033 gross register tonnage, 996 net register tonnage, which are measurements of capacity or burden (a volume unit). Carvel-built on Swedish oak and pine wood, and copper sheathed outside, an outboard protection against woodworm. Paid-up Policy 229/1876. It was insured at the estimated value of 200,000 kr, a fortune. Securities/shares as of 1895 held by whole-sale dealer and shipper Per Adolf Engström, Gefle (Axel Söderström's former classmate) one-eighth; Mrs Katarina Charlotta Engström, widow of the late Lars Petter Engström, the father of Per Adolf Engström, five-eighths; the ship builder O. A. Brodin

IV. The Voyage in Context and Retrospect

one-fifth; and harbour captain Carl Annerstedt, Malmö, one-twentieth. The shipmaster was never among the number of shareholders.

Carrying the Swedish–Norwegian Union colours, a universally recognized signal was also assigned to the ship, HNJB, showed at masthead with signal flags in a row (see page 40). The colourful flag symbols used for this purpose, when hoisted together in line, each had alphabetical signification according to a code of signals, whereas used alone, one by one, they meant:

Fig. 21a. *H= I have a pilot on board.*

Fig. 21b. *N= No/Negative.*

Fig. 21c. *J= Semaphore message to follow; (or, in later days 'I am on fire and have dangerous cargo on board: keep well clear of me'; or, 'I am leaking dangerous cargo').*

Fig. 21d. *B= Explosives; and so forth; all such flags also corresponded to messages in the Morse code—when that invention was introduced and became common later on; telegraph was never available on this ship. On the other hand, the crew was dexterous in hand flag semaphoring!*

Hence, when the *Atlantic* set sail on its maiden voyage on 10 July 1876, its nationality and identity would be recognized worldwide when it hoisted the five flags. They did so now, heading for Ljusne in Hälsingland to load timber, destined for Bristol, United Kingdom. First mate was Carl Israel Svedberg (1842–1904), formerly the skipper of the Gefle brig *Baltic*, which had been sold that year.

13. The First Series of Voyages

The maiden voyage went well, but on an ill-starred journey between Hamburg and Cardiff in February 1879, the *Atlantic* collided with the British brig *Anthes* off the Lizard Peninsula, Southern Cornwall, in the United Kingdom. The brig foundered. Five men in its crew were saved ashore, whereas four remained missing [Nordenberg, 2014, 46]. Back to Gefle on 20 June 1879.

19 July 1879: departure from Gefle in ballast, bound for Australia via the sawmill at Svartvik, Sundsvall. First mate was Gustaf Blomberg (1853–1936), who recalled: 'In the spring of 1879 I was appointed chief mate of the then largest Gefle ship *Atlantic* for a journey Gefle–Svartvik–Australia. The captain, Axel Söderström, was an exceedingly fine, agreeable, and considerate person, with whom I experienced a short but very pleasant time as officer' [Blomberg, 1934, 21]. They made the distance Svartvik–Port Adelaide in ninety-six days! Then they went Port Adelaide–Bordeaux with wheat cargo in ninety-three days, although the main mast was broken and secretly makeshift repaired in Port Adelaide to acquire a sailing permit. This successful trip is described in Gustaf Blomberg's memoirs of 1934, quoted above. The *Atlantic* returned to Gefle on 23 July 1880.

23 August 1880: new signing on in Gefle. Out from the harbour on the 28th, loaded with 66,390 cubic feet wood from Stora Kopparbergs Bergslags AB, bound for Port Adelaide, Australia.

On 15 September 1881, departure from Sundsvall with timber, via Wifsta Wharf, again destined for Port Adelaide. The captain joined the crew in Gefle on 5 September, probably after having touched base, for once. They passed Øresund on the 22nd, at the same time as the

Gefle barque *Monarch* (captain Magnus Wretman[7]), which likewise had loaded timber at the *Wifstavarf* and was heading for Port Adelaide. The *Monarch* arrived at her destination on New Year's Day 1882, the *Atlantic* two days later. Along with the former arrived the full-rigged Gefle ship *Callao* to the port, also carrying Swedish timber, under the command of Johan Petter Påhlsson; and moreover, the Gefle barque *Tropic*, navigated by Svante Kellner, arriving from Cape Town, South Africa.

What is more, the brand-new Gefle barque *Solid*, navigated by Pehr Bernhard Waxin, one of Söderström's numerous relations among skippers, arrived to the port on the same day as the *Atlantic*, on 3 January. Imagine the crowd of Swedish ships in Port Adelaide then, and the jovial reunion of the five captains—all old comrades and almost of the same age, at the dockland café in the Australian summer heat of New Year, on the opposite side of the Earth!

Captain Söderström signed off in Sundsvall in August 1882.

Disbandment, nay! The captain was not a homebody and kept away from homeport for years. Ocean bound, he was in his element as navigator (Figure 19).

In no time at all, on 19 August, there were general sign-ins at Sundsvall; and once more they were bound for Port Adelaide with timber and woodware. They landed at the destination on 16 December 1882. In September 1884, the ship covered the distance Söderhamn (Sweden)–Senegal in thirty days at a new record speed [Nordenberg, 2014, 46].

Captain and crew signed off in Gefle on 2 September 1885, i.e., after three years voyage with a very complex itinerary—in parts of which at least one of the Söderström daughters participated according to family tradition. This is not substantiated by the incomplete documentation of this extensive trip, but may explain some of their later recollections, as can their subsequent travels. And, perhaps even more, why Maria does not tell much about the ship and the crew in her diary of 1885–1887—as the setting and milieu was by then already familiar.

---

[7] Model for the novel hero John Grogg, created by Captain Georg von Schéele, who earlier figured in this book; also well-known to the Söderström family [Schéele, 1889].

She reportedly appeared with her father in the Port of Drammen at the Norway coast in 1884, a base for the captain's business on the next journey.

After the preceding adventurous tours, the ship was on repair in the first part of September 1885 at a dock in Gefle, having been overhauled and coppered there. Thus, it was then prepared for the journey described in Maria Söderström's journal of travel, which started in late September!

14. Setting Sail

25 September 1885: Signing and embarkation in Gefle, commander on the bridge, captain Axel Söderström. Since his daughters had passed the final eighth grade in the pioneering Själander Girl's School, the captain takes the opportunity to invite them to take on as guests and apprentices for educational purposes, and to be able to see them on a daily basis again. After his nonappearance at home as a deep-sea sailor during their childhood years, there is much to make up for.

Accordingly, his daughters take on, Maria (b. 1867), and Emmy (b. 1868), and the former keeps her unique diary. Both take down memoranda and write letters and were to tell about their experiences from the seas for the rest of their lives. Thanks to this memory bank of observations, they meet us on board, as it were, yet without seeing us, but we see them. And by magic luck we can follow them in their reports almost day by day and get a sense of the realities on an old sailing ship.

First mate up to 21 November 1887 is the trusted and cherished Pehr Gustaf Elfström, who was then replaced by Carl Ludvig Johansson. Second Mate is Johan Leonard Andersson Clase, the son of the captain's sister Johanna and his former boss Anders Clase. Departure from the sawmill and its stocks of well dried timber and deal in Skutskär, Upland, on 7 November—having then again received an assorted share of felling in the domains of the world's oldest forest industry and mining company, founded in the Middle Ages. Finally leaving Gefle and Sweden behind now to meet the demands of the Australian building trade!

However, the market is dipping, competition topping, the economy is in recession, dearth of means of livelihood propel mass migration to continents overseas. The year 1885 is one of worst and most strenuous in the annals of trade and shipping.

The historical setting is special, and we have eyewitnesses onboard. Thanks to Emmy's recollections and other available sources we can reconsider and examine this trip in some detail, and from another point of view, adding further aspects to the picture. The full context is in focus.

15. The Take-off in 1885

The moment was fraught with expectations, uneasiness, and emotions.

Departure time, the ship's bell chiming! Young and old were in the harbour to say goodbye—neighbours, schoolmates, friends, spouses, children, stevedores, dockers, loafers, idlers, senior citizens, and others interested to watch the formidable ship cast off—while others were still busy boarding with goods and baggage and hopes, combined with the delight of being decently employed.

On the jetty and the embankment there was an assembly of family members who shared some of these feelings yet took an opposite view on the other side of the dark water. It was a day of rejoicing and wet handkerchiefs, shouts and acclamations, laughter and tears, smiling as well as inconsolable faces. The ship's bell chimed again, the gangplank was removed with force, the captain roared as a lion to the crowd of youngsters flocking at the stone edge, rebuking them to keep clear of the quay! The harbour pilot had arrived for the tow; all provisions checked! all loading done! the seamen happy to get away and have an end to the heart-rending leave-taking scene, what a relief! The gravity of the situation did not escape anyone, though. To some, there was a lurking fear of a definitive parting, an irrevocable departure—a point of no return—on both sides of the withdrawn gangway, the now broken connection between the two worlds. It was a deeply affecting moment for all that were thus left alone; not unparalleled, but so often repeated in seaports and railway stations worldwide, a farewell scenery of loss and the onset of endless longing…

Putting off from land, chief officer Elfström's wife and family waved goodbye, the seaman leaving spouse ashore alone for years, as always, with a vow of chastity. Other married men in the crew were the able seaman Johan Gustaf Bergström and the steward & cook Carl Östling, who took leave of dear ones at the wharf. Boarding complete, Bergström was oldest in the crew of seventeen men, born in 1838; ordinary seaman Knut Axel Boman the youngest, born in 1869. Thus, they were all between 16 and 47 years old, and the girls not the only teenagers on board; Gustaf Theodor Lysander a low-paid deckhand was the same age as Emmy (Figure 11).

Besides the first mate Elfström and the cook Johan Axel Waltner (b. 1864), the steward Östling was an important contact and employer to the young women, and they got along very well. Born in 1848 he was an experienced man with a number of assignments, and a central and popular role as chef and headwaiter, organizer and staff manager. He had been on the seas for a while and picked up foreign recipes as well as acquired considerable proficiency in his different duties. Above all, he was ready-witted and a lucky fellow. He had sailed with the resplendent ship-rigged merchantman *Indiaman*, cleared inwards at Melbourne on 20 January 1872, where he jumped ship and, in this way, escaped the ship's foundering at Cook's Bay, Easter Island.

Having a good way with young hands, as well as animals, living or dead, the supply of primary produce was nevertheless limited before the age of freezers and tinned goods. They carried salted pork, pickled herring, dry beans, root vegetables including potatoes and turnips, onions, smoke-dried lamb, air-dried codfish. On the whole, this was the provisions of Viking sailors a thousand years ago, including the Icelander Erik the Red, the expat outlaw, who discovered Greenland on exile, and his son Leif Eriksson, who anticipated Columbus when he crossed the Atlantic further westwards in his sailing vessel—the sole outcome of which was merely rudimentary genetic streaks of American Indian heritage in the otherwise heterogenic Icelandic population, though.

The captain—who in his youth wanted to become a physician and dabbled with medicine, and now had sole responsibility of the health of crew and passengers—prescribed a citrus fruit a day for all, as preventive medicine against scurvy. That was a matter of daily routine which his daughter Emmy was to follow for life on land, as were her children and grandchildren in a later era.

16. Close-up Picture of the Indispensable Chief Officer

Pehr Gustaf Elfström oversaw and monitored all work on the ship and kept it maintained and going, while the captain preferred to contemplate at the wheel. Elfström died on a voyage between Le Havre and Cardiff with the schooner *Frey* as commander of that ship on 28 June 1890, disembarking with death. Then all was well on board, only the captain missing.

His previous meteoric career from assistant deckhand to mate and shipmaster had not been without trials—such as when he departed or rather escaped from the ship *Magellan* in Callao on 10 March 1872, missing his ship and finding himself left astern west of Lima in Peru, on the South American west coast!

Sea captain Söderström trusted him completely, and signed him at a monthly pay of seventy Swedish crowns—highest of all wages on board—for his marked competence and good manners, and he was always popular with the crew and much liked by the ship's girls, whom he looked after.

Elfström passed away at the age of 35 years 11 months and 8 days, leaving a small number of surviving relatives: his bereaved wife, now widow, Johanna Christina Elfström, born Berglund; his mother, the widow Brita Helena Elfström; a 10-year-old son, and a few personal belongings according to the Estate inventory: A gold ring (his betrothal ring); a pocket watch made of silver; six teaspoons in hallmarked silver; and, in nickel silver: six tablespoons, six teaspoons, one sauce spoon, and one castor spoon—his only assets!

His life was typical of sailors: basic education, apprenticeship, adolescent years on ships, short stays at home, and looking for a life companion in the port of registry; occasional reunions with family, all life on the sea, long travels with hard work year after year, little private means or property, no fortune.

Only necessities and necessaries were stored in seamen's chests—including, if they were lucky, such things as pipe and pouch, or an oblong snuffbox of metal with bandanna; a flask, boots, a seaman's sheath knife, a pair of scissors, cutlery, blanket or travelling rug; a worn old book or a cabinet photograph; phosphorus matches, comb of bone or brush of pig's bristle, a toothpick, some clothing along with the personal mariner documents; and in good times: toothbrush, razor, towel, some foodstuff from mother at home.

Along with this, sailors carried considerable seamanship and strength, and a large fund of knowledge and experience. Elfström was a man of passion, sentiments, and learning, an efficient mate full of energy and nerve.

His monthly salary, called *hyra* 'rent', because the berths were formally hired, always paid in cash by the master—who was the manager, public authority, cashier and staff department in person—was comparably fair at the time. 70 kr in 1885 could buy as much goods and services as about 5,000 kr in 2021, measured by the consumer price index, i.e., something around a 500 note in euros or US dollars. But that is not saying very much.

17. The Living Conditions of the People

To get some idea about what this meant in the practice of everyday life then, one should remember that his monthly payment of seventy kronor answered to the annual salary of a maidservant, an agricultural worker, or a farmhand. At a cafeteria or snack-bar in a Swedish seaport, you could be served three meals a day with coffee or small beer at 20 kr per month. You could get a large loaf of good bread for one krona and buy a *gateau* for two kronor.

In comparison, crofters and many other day labourers, earned one krona a day, as did the old shipyard carpenters.[8] The one advantage

---

[8] By the way, the low wages in the Gefle harbour and the shipyards made many of their employees sooner or later to seek their fortunes abroad. Among them crane operator Joel Hägglund (b. 1879) who as Joe Hill became a famous protest singer and working-class hero in the USA. The Gevalian deep-sea sailor and helmsman Claes Johan Mineur replaced Joseph Conrad as commander of the stern-wheeler *Roi de Belges* in the Congo, the riverboat in Conrad's *Heart of Darkness* [Henricson, 1990, 85, 91].

for sailors—as for maidservants and nursemaids—was free board and lodging, but they had to keep working-clothes. Seaman's blouses did not exist but in comics and children's wear.

Laundering was a rare opportunity, and people spending their lives on sea lanes and in waterways were grimy. Some spent a coin or copper at bathing places in foreign towns and cities, where there were other enticements, too, and smart businesspeople to boot, alluring to a jaded working party in unbridled, much longed-for leisure in search of adventure, in environments where they were anonymous among perfect strangers. Wages were sometimes sent home, but mostly spent in seaports, where there were often favourable exchange-rates and low prices. Each Swedish city had a savings bank, a novelty then and a fairly new feature in the countryside, but a common sailor with a passbook was almost unheard of.

For those back home on land, housing was expensive, primitive, insanitary and poor—for town-dwellers as for factory workers and agricultural labourers. The upper crust bought blocks of flats and constructed great new buildings with apartments to let for the rising city middle class. In 1885, the large rental property Överkikaren 16 in the Maria parish at the Lock in central Stockholm, with a plot area of 358 square metres, was sold for 56,000 kr, and rent racketeers had palmy days before national rent control was introduced after the turn of the century.

In the nineteenth century, Sweden had a very low standard of housing, it was cramped for space, many people were living in overcrowded conditions, unhygienic and unhealthy. In the middle of the century, the majority of Sweden's population lived in so-called single cottages or couple cottages in the countryside as in the townships. In the duplex building was a vestibule that led to a living room on one side and a guest room on the other side. The latter was only heated when it was used. Behind the vestibule was a chamber that was utilized for several different purposes, often used for storage, and as a cloakroom. In this duo lived ten to twenty people, where everyone slept in the same room during the winter to keep warm. In this chamber, everything was done, from eating, sleeping, breastfeeding the children to doing chores of housework. The kitchen was the most frequently used space in the home and was therefore the dirtiest place in the house.

During the latter part of the nineteenth century, industrialization broke through, which changed the conditions for housing considerably. The migration from the countryside to the cities led to housing shortage among the workers. They were poorly paid and could not afford to rent spacious apartments anywhere. A room and a kitchen were what some of them could afford for their nearest and dearest, and it was not uncommon for two families of a total of twelve to fourteen people to share an apartment. Alternatively, the families who had a room to themselves with private entrance, were compelled to welcome one or two bachelors to sleep on the floor or the sofa to get along. Unemployment for the workers almost always meant notice to quit and eviction, as they could not pay the rent, and landlords and house-owners did not wink at anything. Single men often lived as lodgers with care-of (c/o) or Poste Restante addresses, or in bachelor hotels.

No wonder that joining the merchant fleet, becoming a mariner, and to see the world, was beckoning during these circumstances, and a beacon of hope to thousands! What is more, the availability of low-price merchandise abroad and on trans-boundary waters, seemed attractive to many. To exemplify with a very small everyday article: safety matches, a Swedish patent of 1844, these were much in demand those days, and a successful export commodity. Few sailors could afford them, though, but kept with the old phosphorus matches available in every port and of universal use, often free of charge at coffee shops and beer houses as promotional items to the customers. Nor would any one of them buy the tobacco brand Gefle Vapen from P.C. Rettig in their homeport Gefle, which was popular at fashionable hotels. So, you would find few domestic trademarks in a contemporary sea chest once the owner had left his nest.

Bringing hardtack and mead was ordinary but old-fangled; in comparison, snuff, chewing-tobacco and coffee were relative novelties, real New World items. Turning to the benefits and secrets of officers and chiefs, bringing bowls and bottles of the sweet and syrupy Swedish Arrack Punch, with 30% percentage of sugar and stored with the master to be served on ceremonious occasions, was a local ethnic custom, peculiar to these northerners. Some maintained that the origin of this eighteenth-century liqueur was Indian. Others thought that the invention depended on the sailing East India Company import of arrack from Java.

There were famous and popular brands, such as the Cederlund's and Grönstedt's, introduced by wine-dealers in Stockholm, and there was a local Gefle brand too. Principal seaport hotels used to have 'Punsch verandahs', where directors, proprietors, shipping agents, and tradesmen of the bourgeoisie used to sit and enjoy their quiet cigars in the evening, attended upon by chic waitresses, with whom these wealthy men not infrequently married—the few women they saw in their completely male-dominated business world! This happened to some of the most successful entrepreneurs and factory-owners in Gefle, and according to town gossip it was nuptials to mutual advantage.

As we have seen from the diary, there were entertainment and festivities on the *Atlantic* too. In the saloon there was a piano at the wall, not on the regular wheels but nailed to the board as other furniture. In good standing, too, and well-tuned, to the young ladies' delight, who were practicing on the keyboard. It was not only meant for their own enjoyment, however, but for their education, and moreover for entertainment on board, besides accompaniment on special occasions, the latter usually of sombre nature!

In the captain's cabin sat a wooden desk and his armchair, which survived him and still is preserved; his ledgers and nautical instruments, bookcases and shelfs transfixed and secured, panelling and wainscoting in stained and lacquered deal. His weaponry and ammunition were stored there in fixtures along with insurances, travel documents, clearances, agreements, ship's lists and the money, along with charts and log—and his strong sweet liquors, more exclusive than a pirate or common decent sailor's rum and spirits, but not the least bit healthier. The stern captain with enormous willpower was a sensitive and sentimental man who did not begrudge himself nor the crew well-earned *joi de vivre* at opportune moments, and at special occasions the golden 'Punsch' was served to all the crew.

18. Contemporary Celebrities and Events

In India, in the year of 1885, the Polish seaman Józef Konrad dispatches his first letters in English before setting sail with the clipper *Tilkhurst*, bound for Singapore; thereupon navigating the barque *Otago* from Sydney to Mauritius. He is then about to shift both career, name, and language to become the foremost British writer of sea novels, all based upon his own experiences from the *Torrens* and many other ships.

The frigate HMS *Vanadis* returns to Stockholm from its spectacular research expedition round the world, alluded to in Maria Söderström's diary. The year 1885 is also a great and crucial year for world-famous painters such as the Russian Ilja Repin, the Dutch Vincent van Gogh, as well as the Frenchman Paul Cézanne.

In France, Émile Zola publishes his *Germinal* about strike action, the current and at the time topical issue termed 'strike' in actual fact being partly of maritime origin, connected to 'striking sail' and the well-known tradition of 'refusal to work' and mutinies on the sea, incidents quelled also on the *Atlantic*.

Furthermore, in the realism of the contemporary world of fiction, the ill-famed Georgia-registered barque the *Lone Star*, homeward bound for Savannah in 'The Five Orange Pips' by Sir Arthur Conan Doyle, with its Ku Klux Klan captain and mates wanted for murder, is lost at sea during the severe gales in the autumn of 1885. The only trace of the vessel is a ship's sternpost marked 'LS' sighted in the North Atlantic.

Symphony No. 4 by Johannes Brahms is premiered in Germany, with the composer himself conducting. More importantly, Gilbert and Sullivan's new operetta *The Mikado* opens in London, the play that was produced in Melbourne in the year to come with Emmy and Maria Söderström in the captivated audience, a performance that was crucial to them and praised in the journal.

In the world outside theatres, concert halls and auditoriums, and beyond book stalls, lounges, and damsels in distress, wars are raging between France and China, Serbia and Bulgaria, and between the British and the Burmese empires.

The mass migration to America and Australia peaks.

On the sea also, in the smaller format, Captain Joshua Slocum, later famous for the first single-handed circumnavigation, finds himself unsuspectingly delighted by his barque the *Aquidneck*, 'the nearest to perfection of beauty'. As it turns out, the ship is on its way towards a tragic finale, though. The crew is struck by cholera and smallpox, a mutiny aboard breaks out, in which the master shoots two men, and the ship is finally stranded on a sandbar and wrecked in 1887. Whereupon the man in command, who was the owner of the ship, reaches a nadir of fortune—and becomes the well-known author of a sea diary written on a 36-foot-9-inch sailboat.

The successful Swedish mercantile marine, essential for the nation's exports and imports since the Hanseatic League in the Late Middle Ages, reaches its acme and culmination in the major development of international shipping that took place during the Victorian era, 1837–1901.

The growing intensity of worldwide communications has a variety of outcomes. The famous mariner and collector of lacquer, bronzes, textiles, ceramics from Asia, Rear Admiral and Flag officer Sten Ankarcrona (1861–1936), first gets a taste for Asian Art while serving as *Enseigne de vaisseau* for the French navy *La Royale* from 1885, when his frigate sets off for the Far East. The outcome is a discovery of the beauty of Japanese artwares, handicraft, and cabinets, which henceforth embellish Swedish home furnishing. All destinations have repercussions on the ports of departure, as all foreign influence tends to become interactive. Interior decorations in the homes of occidental deep-sea sailors, likewise, turn oriental. The world changes reciprocally.

Modern breakthroughs in Scandinavian literature sets in, with a focus on female emancipation, a main theme that universally booms in the latter half of the century with all its dissatisfied women longing for freedom—reflected in famous literary works by men, from Gustave Flaubert's French *Madame Bovary* to Leo Tolstoy's Russian *Anna Karenina* and Henrik Ibsen's Norwegian drama *A Doll's House*. In addition, this theme corresponds to a change of mindset that takes place in the society, which is progressively pictured and influenced by the increasing number of radical woman authors and travellers, and their attempts to achieve freedom and equality [Hallengren, 2001].

As other young city women of the rising educated middle class in this emancipatory period, the sisters Emmy and Maria, who have left the liberal Girl's school, avoid stays and the special undergarments of older high fashion. As can be seen from their dressed-up appearance on the photo from the *Atlantic*, showing their finery, they wear common long wool dresses fitting loose and no tight garments. Fashion is now turning back to the national and Old Norse cultural heritage, to natural and comfortable clothes of ancient folksy origin and style, supported by the rising handicraft movement, the defenders of domestic arts and crafts. In 1885, a 'reform dress' is introduced in Sweden with a similar political program for the gowns of haute-couture, aiming at a bodily liberation of upper-class women, relieving them from corset and bustle—the latter feature an excessive phenomenon poked fun at in Maria's diary. Consequently, we can discern the turn of the times and current changes of fashion by looking at the sisters, even if they have neither bought nor chosen their outfits. The ornaments and buckles on their dresses can be traced back to the Bronze Age, whereas long trousers for women still belong to the future.

Christina Nilsson, the world-famous female Swedish soprano who inaugurated the Metropolitan in New York and is mentioned in Edith Wharton's *The Age of Innocence* and Leo Tolstoy's *Anna Karenina*, in 1885 gives a farewell concert from the balcony of the Grand Hotel in Stockholm before an audience of 50,000 people crowding the street below at the quay. A record attendance. Panic in the jostling throng of people breaks out and several are injured or killed, some trampled down and others drowned. Nineteen fatalities—all bodies brought into the hotel to the singer's horror and despair: 'My fault!'

Her compatriot, colleague, and compeer among woman singers, the terrific Jenny Lind, 'the Swedish Nightingale'—by then a household name in the New World as in the Old—passes away two years later after a life as a world star. When Ralph Waldo Emerson, the leading American philosopher, writes his essay on 'Success', Jenny Lind is one of the three outstanding and universally known women he alludes to—the other are the authors and radical freedom fighters Harriet Beecher Stowe and Margaret Fuller Ossoli.

On 7 November, 1885, when Maria Söderström's *Notes during the sea voyage* begins and the travel round the world starts, a vast distance on the ground is simultaneously overbridged by the Canadian Pacific Railway, extending across Canada by reaching British Columbia that day.

On the sea, the weather is not only hazardous in contemporary fiction. A record-early snowstorm strikes England in September with fatal consequences, and in October a hurricane hits Labrador, the coastal region of eastern Canada, destroying eighty fishing vessels and killing seventy men.

Keenly aware of reported weather conditions and wrecks, as well as the challenges in the Nordic winter month of November, the seafarers in our book salute the surprisingly warm nice weather when they set off as a good omen. This is the reason why the captain exclaims with satisfaction:

—'Unparalleled weather in the month of November!'

19. Emmy's Recollections

With an embalming light breeze in the sunshine with a temperature over ten degrees Celsius on the shady side of the ship, they sneaked down the west coast of the island of Gotland in a slow pace as if out sightseeing. The ancient, fortified city of Visby was something to see in this slow fashion, appearing at its best in the warm autumn light. There was no Gotlander on board, but all were born and raised in old ports and fishing villages on the east coast, except Lysander who was born inland in Scania. The passage was familiar, and some of the officers and crew knew their history book. The bells chimed in the medieval Church of Saint Mary, which Maria observed. The young women at quarterdeck descended to portside rail, and they soon got company.

The ring wall is something like an ancient medina or kremlin. Visby was fortified like most other cities in antiquity and medieval times, by the sea or built on a hill, from Alcúdia to the city of Rhodes or the ancient Lindos. Defence measures turned chief towns and capitals into citadels and castles until fiends and foes became intangible as spectres and more difficult to keep out.

In the field of vision appeared the ruins of the St. Nicolai church along with its Dominican convent, where the fourteenth-century monk, Peter, according to the sayings, fell in love with a foreign nun and through his writings became the first Swedish author and love poet. The church built in honour of the Saint Nicholas is one of the most conspicuous and well known among the ruins. The captain added that Saint Nicholas is the patron of sailors—and of fishermen and merchants, the mate chimed in, 'so he is with us. Early stories tell of him calming a storm at sea. We need that.' 'Also a guardian of coopers,' carpenter Blomgren added. (And prostitutes, they wouldn't tell or wouldn't know.) The saint had many protégés.

Next day they returned to reality, and all were busy at work, except the girls who lay seasick in the rolling sea. Officer Elfström's joke that he only became sick in calm weather, proved true. The sailors were used to hard struggle and wanted to get forward.

Emmy remembered the feeling of losing control, not knowing where her things were, shaken to the bone and nauseous, put out in a state of confusion and dizziness. Then she was finally cured by standing outside on quarterdeck facing the gale and the spray of saltwater, and focusing on the bowsprit and the horizon, following the movements of the ship with her eyes, and not on any account retreating to the sleeping berth. She also learnt the necessity of eating well during the meals to avoid indisposition. That day in the Skagerrak was her only fit of sudden illness during her years on the sea. In the same waters the doom of the *Atlantic* was to be sealed in the future.

On this outward passage, then nobody blamed Johan Axel Waltner, the eminent cookhouse chef of the day, or Carl Östling, the steward, who served the meals and waited on them at the table as a noble butler—a settled man now far away from matrimonial morning coffee in bed. The herring and dill potato dishes with bread and butter, milk and porter, raspberries with cream, were spic and span, as were the still immaculate linen. The cordiality and kindness warmed the cockles of Emmy's heart. Outside, it was getting increasingly cold and windy on the seaway below the North Sea, where, off the west coast, the faint nocturnal blue band of sea-fire in the wake after sunset had been a sight for sore eyes.

Coming into the warmth between the Tropic of Cancer and the Tropic of Capricorn was indeed delightful, and the crossing of the Southern Atlantic to the 34th latitude went smoothly. However, after the left turn off Uruguay, and on the long, drawn-out trip towards the Southernmost point of Africa, the rocky *Cabo das Agulhas* 'the cape of needles', through the Cape of Good Hope and beyond in rough sea, and on to the passage through the Indian Ocean in the extreme South, was no simple preliminary. It was cheerless and inspired sombre thoughts. Her sister stopped keeping diary and withdraw.

Ascending the poop, and later moving ahead before the mast and taking up her stand at the bow, Emmy got a view of the slowly breathing sea, that had turned from a rippling turquoise, blue, and mauve to black under a sky of fluffy paratactic clouds on the move, the wind silently whispering in her ears 'evermore, evermore,' and remembered the report of a spring tide in lower deck, tales of seamen who were washed overboard, and recalled the lookout in the crow's nest who fell from the mainmast into the Abysses, and how a captured boat had signalled that all was well on board bar the captain gone. When the haze dimmed the outlines of ships and skerries in the deep of darkness at nightfall, she thought of the legendary ghost ship the *Flying Dutchman*, navigated by a captain who failed to pass the Cape of Good Hope, and whose fate is to stay on the seas forever.

The crew was happy to spin their yarns in her company, and she was all ears. When they were all very busy, she at times had the opportunity to hear some of them sing shanties, old maritime work songs.

Travelling along the curved Australian South Coast, passing through the South Australian Basin and further southeast, there was much to see. On approaching Sandridge (Port Melbourne) in bright daylight, slightly more than a hundred nautical miles from the city, they off the shore of Port Campbell at 38°39'57"S 143°06'16"E passed the eroded limestone stacks named the Twelve Apostles. Then the crew for a moment flocked to the port side, making the ship tilt a bit more. Maria was stunned by the vision of the apostles, whereas Emmy in a high voice, making herself heard above the surfs, spoke about the beauty and wildness of the cliffs and questioned the number of the disciples. Both of them feared this approach, nonetheless, which was known for its perils and numerous wrecks, and the helmsman kept clear of the shores. The sea was safer than land.

When shore-based, Melbourne was to become their home in 1886–1887, the point of departure and return. It was a crucial point too, from which extended excursions to North America and the central Pacific started. These voyages brought them along very different sea routes and included many short touchdowns and anchorages on outward and inward passages, and they saw many different people on the way. They encountered native Americans, Polynesians including Maori, Melanesians, and Micronesians.

Emmy was surprised to see a couple of very dark-skinned people of the Melanesian islands with blond hair and was told that it was a result of earlier European landings in New Caledonia, but today we believe that a unique gene is the reason and not encounters with visitors of the past.

The old thin Papuans with lined faces and handsome complexions which the captain and mate communicated with on temporary anchorage in the lengthy repeated passages through the Pacific, on deviating routes and in shifting weather conditions, she thought looked like old wise men, and which in an amiable way, and silently amused, answered her naïve questions about the taste of human flesh—experienced seamen and harbour people as they were, likewise in the intense international trade, doing business as themselves.

However, the only indigenous peoples Emmy and Maria got to know by personal acquaintance on a daily basis in the 1880s were the young workers from Niue and Aitutaki who brought them on exciting nocturnal fishing trips during their prolonged stay at the phosphate-rich Malden Island. That is, the atoll in Kiribati with its breeding colonies of seabirds and strange prehistoric ruins; later laid waste by atom bombs when it became a nuclear test site.

Emmy remembered New Zealand as the most enjoyable and beautiful place on earth. In point of fact, it was quite similar to the nature of Scandinavia. Emmy recalled a short stop at Christchurch on the South Island and attending service in the later well-known Anglican cathedral in the small town, still under construction, but with the magnificent nave and tower already in place. She was sitting down in the pew and waiting in silence. The edifice was built by stone due to the lack of timber on the island, the product that the *Atlantic* was transporting to this part of the world and was marketing there.

It was church service then, with the old dean Henry Jacobs officiating, with British hymns and readings from the King James Bible. Apart from the language, everything was familiar and reminded of home, even the weather and the atmosphere. Emmy remembered New Zealand for its nice climate, overall so similar to the Swedish–Norwegian ambience and landscape. The town was small, too, and easy to take in, with well over a thousand residents in the middle of the 1880s. When the present author followed in the wake and sat down in the pew in 2009, the cathedral was largely the same, but the hymn book in Maori and an indigenous priest officiating, and the Cathedral Square in the midst of a large city with traffic; and, incidentally, the Lunar New Year was lively celebrated everywhere by the large Chinese population. However, the cathedral in New Zealand's first city, which was founded in 1850, has been regarded as a constancy in a changing world, although it was damaged by earthquakes shortly after our visits—in 1888 as in 2011—and many times more, its symbolic value still estimated higher than the perpetual reconstruction costs.

Karen Wright, a Maori publicist, with family and friends took pains to explain to the present author how different Waitaha (Canterbury) was when the Söderströms were there just a few decades after the British had taken over Aotearoa (New Zealand) and Maori culture was pressed back for a century on. However, she confirmed Emmy's remembrances of women with tattoos, and with satisfaction the friendliness the strangers were met with, the Maori *manaakitanga* or hospitality. Emmy only vaguely remembered one concept, though, the *matariki*, the Maori word for the Pleiades, a constellation so well known to the night watching and stargazing sisters and their father of the seas. She recalled this, when she at an advanced age heard that an astronomer had discovered that people thousands of years ago could perceive many more stars in this cluster, and that all these stars are named by the Maori, which in oral legends harbour memories from primordial times, as do the Aborigines of Australia.

Among buildings in Melbourne still standing, the sisters particularly recalled the impressive Royal Exhibition Building (Figure 23), then renovated and empty when they entered the greatest hall they had ever seen. Furthermore, the number of theatres, of which the Princess' Theatre (Figure 22) and Her Majesty's Theatre still are almost exactly as on their evening visits. Even some of the popular plays and operettas they saw in the 1880s, occasionally occur in the nostalgic repertoires, including the *Boccaccio*, which Maria disliked the most, and the *Mikado*, which was her favourite along with dreamy *Rip Van Winkle*.

Emmy in particular remembered one visit to the theatre, which appeared to change her sister's life. It was a family evening, when their

Fig. 22. *The Princess' Theatre, Melbourne.* Photographed by the author.

friend John joined them at the theatre entrance, the nice captain of the *Gurli*. When they left the theatre after the performance and should depart, he moved closer Maria, and their shoulders met for a moment. 'This evening was fine,' he whispered, or something to the same effect. She replied in a low voice, very slowly, putting stress on each word, looking at his bright eyes: 'Let us see what the future does hold for us.'

It was all still, not even a wind, and there were no more words.

She heard his footsteps in the Spring Street leaving, and saw his suede jacket receding, and turned around to the captain and her sister who were watching her quietly, even tacitly; she was not sure. Brought to the world down under, the nadir of their beginnings, they were not at the end of the road yet, but indeed en route, going even farther, there was no turning back.

IV. The Voyage in Context and Retrospect

Fig. 23. *The Royal Exhibition Building, Melbourne, built in 1880.* Photo Anders Hallengren 28 February 2009.

You can never return from a journey, she thought, since the place you left is no longer there, as in the folk tales where you arrive to the same place but in another time.

This night it was difficult to retire, she afterwards told Emmy, intimating everything, telling all her thoughts.

She had been overturned and went for a long walk through empty lanes and alleys after nightfall. At midnight she found a quiet little yard nearby, where she sat down on a wall in the dim moonshine. The First Fleet had arrived a hundred years ago. They were bringing all their belongings because there were no return. Just like me, she thought. Sailors know that you can never find again the same sea and your hometown as you left it. Nevermore, nevermore.

Moving stealthily back to her room in the wooden townhouse, tiptoeing on the creaking floorboard, she was overcome by fatigue, did not light a candle and say goodnight or her bedtime prayers, and slept without dreams.

V. The Last Days of the *Atlantic*

Fig. 24. *The ATLANTIC resting in harbour. The barque Atlantic was built at O. A. Brodin's Shipyard in Gefle in 1876.* Courtesy of the Länsmuseet Gävleborg Photo Collection. Identifier XLM.A98-34.

1. Nuptials

AFTER THEIR BETROTHAL AT CHRISTMAS IN 1886, and the heart-rending goodbyes in January 1887—when Captain John Ljungberg had to leave Melbourne with his *Gurli* of Gothenburg—Maria Söderström had hoped to see her fiancé again on return to Hamburg and Europe in September. This hope was dashed to the ground. The *Atlantic* and the *Gurli* passed one another, and their letters crossed in the mail. They finally met again however, and were married in Gefle next summer, on 28 July 1888 (Figure 25).

Blossoming of supreme happiness and rosy youth at this elevating moment, the 21-year-old bride in the evening was reminiscent of Vermeer's *Girl with a Pearl Earring*.

Fig. 25. *Maria Söderström as a bride on 28 July 1888.* The Maria L. Hallengren Collection, Sweden. Private possession.

It was then decided that she should move to her husband's home district in the southern seaport Westervik (Västervik), whereupon they subsequently settled down in the two-storeyed roughcast wooden farmhouse No. 157 in the eastern quarter, built about 1882. This was in the neighbourhood where her husband grew up, and his seafaring ancestors had lived since the eighteenth century and where he was to stay all his life on land.

Then again, this relocation did not take place at once, and, fearing a new period of loneliness and longing, she was permitted to embark the *Gurli* and accompany her John on the seas. Hence, after the years on the *Atlantic*, she was now on the *Gurli*, switching from one ship to another and staying on the oceans.

On 24 July next year, in 1889, John's and Maria's firstborn son Axel Erik Douglas Ljungberg saw daylight, apparently conceived on the trundling sea as his mother was.

According to the Census of 1890, they are then finally domiciled in the '*Eastern Quarter No. 157, Västervik, Southern Tjust; Ljungberg, Johan Erik, b. 1852 in Västervik, Kalmar county, sea captain; Söderström, Maria Matilda Charlotta, b. 1867 in Gävle, Gävleborg county; Axel Erik Douglas, b. 1889 in Västervik, Kalmar county*'.

On the *Atlantic*, it was business as usual and all went on according to schedule, unperturbed by private matters. Yet, the captain was worn, his ship the worse for wear, the times hard, and the palmy days gone. However, the voyage, which had started in 1885, had not yet been completed.

Carl Ludvig Johansson, who had replaced P. G. Elfström as first mate in November 1887, was succeeded in this position by Rickard Garberg (b. 1854), who took over from Johansson as chief officer in Liverpool, the English seaport, on 19 November 1889.[1]

In the city at the mouth of the River Mersey, Garberg was first acquainted with the captain's daughter Emmy 'Emy' Söderström, still unmarried but secretly engaged. She was still travelling on a journey and stayed with her father at a hotel in the harbour during the extensive bargaining and recharging period.

---

[1] According to the enrolment registers, he was signed off on 11 September 1891, and enrolled again for *Atlantic's* trip in 1892–1895. After the voyage in 1895, R. W. Garberg was one of the three crew members who were given a testimonial ('Good') by the captain, which means that he was recommended strongly.

The party did not leave the port in Liverpool until 9 February 1890, when the ship was heavily loaded with coal, bound for Argentina.[2] In Buenos Aires, Söderström reportedly checked in at 'Hotel Söderström' as a rule.[3] They left Buenos Aires in ballast about 10 May, sailing for Port Townsend in North America, awaiting order and another timber cargo. They managed to purchase and load fine Oregon pine in Gig Harbor, the bay and seaport on Puget Sound, in Pierce County, Washington.

Then they sailed for Melbourne, Australia, successfully marketing the cargo there. Next, they departed for Port Germein in the Spencer Gulf, the westernmost inlet on the southern coast of Australia, to buy and take in a cargo of wheat. On 24 March, 1891, they set sail from the gulf with the grain, and after many trials and tribulations via Cape Horn reached Rouen, the port on the river Seine in France, on 5 August, i.e., after 134 days.

Captain and crew with satisfaction signed off in Gefle on 14 September 1891. The sailors needed time for recovery, and so did the ship, which was to be repaired and renovated in the dock. The *Atlantic* hibernated in Gefle in winter quarters in 1891–1892 (Figure 24).

However, once on land in his hometown, Captain Söderström rushed to his 'father of the bride duties' and walked his daughter Emmy down the aisle at her wedding on 24 September.

Busy days, and large bills to pay!

First mate Rickard Garberg gives a vivid depiction of the eventful trip of 1890–1891 in his logbook, which offers a new approach to the state of things on the merchantman *Atlantic* and makes it possible to see Maria Söderström's travel account from a completely new angle [Garberg, 1962].[4]

---

[2] The loading berth in the chief port of Argentina was troublesome because it was not yet a deep-water harbour for deep-draught ships, and had no real wharf or dock. This is why the *Atlantic* and other ships lay at anchor in the roadstead. The rising international freight service with armadas of merchantmen on the trot encountered similar technical hitches in other goods terminals on the South American east coast.

[3] At a guess, the *Ribera Sur* on the River Plate, the 'south riverbank', which is still a decent tourist class hotel. The Swedish surname means 'south stream'.

[4] In 1889–1895, Rickard Wilhelm Garberg (1854–1942) was the chief officer of the *Atlantic*. The captain gave him a good service record. In his teens, while he was a deckhand on the *Sophie*, Garberg had met with the 'kind man' Axel Söderström (and

2. Mate's Memories

Recently married but penniless and currently unemployed, Rickard Garberg finally could not resist the offer of the shipowner Per Engström to step in as first mate on the *Atlantic*—and leave wife and hometown to get additional years on the sea in his personal record. An agonizing decision for him to arrive at, yet he had no choice. People reproached him, pointing out that sailors should not be married at all, and moreover that it is irresponsible to marry if you are not in funds!

Without further ado, he went to Liverpool to meet up with captain and ship to obtain the means of livelihood.

Having reached the English harbour by night, he found the sombre ship at the quay-berth 'smutty, grimy, and forbidding'. In two minds, he finally hauled his seaman's chest and bag of clothes over the rail. Not a living soul was to be seen, and nobody was there to lend him a hand. He only got so far before the scenery was put in a completely different light when he realized that the appearance of the ship was due to the fact that it was actually loading coal. Everything was sooty.

Bedtime, then!

When I approached the cabin door, I was met by a large, shaggy hound, and I realized that I had to be careful. His countenance was threatening and his growl ominous. When I addressed him in his own language, that is to say, English, he did not precisely light up, but he came along and allowed me to pat him. This way I had made a friend from the beginning, for a long time my only one.[5]

---

the latter's nephew Pehr Zellinger). Out of consideration for survivors, some names were changed when Rickard's son Sven edited his father's memoirs or yarns. Thus, the *Atlantic* was renamed 'Chiquita', and Söderström was called 'Ström'. Nevertheless, captain Söderström trusted his first mate more than his nephew Mr Clase, the second mate since 1885.

[5] This is the dog Harriet so affectionately mentioned in Maria Söderström's diary from the trip in 1885–1887 and was so dear to the sisters. She was a Saint John's water dog, now extinct, belonging to a line in the ancestry of the Newfoundland

The room in the weirdly quiet ship received him with the same reserve; it was 'dark and cold as a bolted ice cellar'. No welcoming stove made the new mate come into the warmth, and he found that 'on the whole there was no heating on board'. He spent the first night in the berth dressed in fur coat, high boots, and jerseys. He was alone, and passed the entire guard on watch.

Next day, he learnt that the crew was paid off; that the maintenance of the ship was entrusted to the second mate who was a drunkard spending most of his time in the city; and that the master 'had moved ashore as well and stayed at a hotel in company with his daughter [Emmy], who had come from Sweden to keep him company during the time of loading'.

The fact that all of them had escaped the floating cold store was not surprising, but why 'this inhuman temperature aboard'? And what about its guard? 'The captain explained to me that he expected the vermin to perish in the cold. As it turned out, this did not work but at last the dog died'.

The new chief officer also found out that dock workers and stevedores were in charge of all loading and did the heavy work, while care of the ship was neglected by the inebriate second mate. Rickard insisted that the second mate—Mr Clase—should be replaced as he was the worse for drink, and the captain acquiesced to this request as he found it justified. The man was accordingly dismissed, although

---

dog and the Labrador retriever. They all stem from the same province in Canada and are remarkably good swimmers. Fishermen setting out nets have used them as helpers for centuries; they have served as lifeguards; and sailors early found their use as watchdogs and pets. St. John's is the capital of the province Newfoundland and Labrador. As this is the legendary *Vinland* of the seafaring Vikings of Iceland and Greenland, there are speculations that the Saint John's water dog arrived with them. In the North, dogs always had a higher status than in the South and the East. They have had a special symbolical cachet among Northerners, as when the famous national chronicler of the sixteenth century, Olaus Petri, stated in his list of ancient Swedish kings that King Attil conquered Denmark and enthroned his dog Racka as the king there. To the national legend also belongs the idea that Harriet's ancestors arrived in Newfoundland with the Viking Leif Eriksson (c. 970–1020) and that they are derived from the now extinct Swedish Dalbo dog.

with regret, since he was a close relation of the captain—a nephew, and an old workmate.⁶

> As would be clear later on, the captain was a real gentleman, but he took no heed of the deplorable condition of the ship, but left it to his new chief officer to take care of everything. He blindly believed in me, and this probably did me good, cheered me up and strengthened my confidence.

Large-scale repairs started. There was ample time. The new man in charge took up his stand on the ship and stood his ground. The watchdog got a companion on board. Soon the company of a second mate, too, an able Swede of twenty-five years, from the cathedral city of Skara, 'along with some rough customers to keep a check on'. In December, the ship was in shape and the berth full. They were bound for Buenos Aires and were ready for departure, weather permitting. A gale of south-west wind stopped it, and the obstacle was from experience expected to last for five weeks. This forecast was not exaggerated.

During these circumstances, and in the long period of waiting, the main task was to maintain order, keep runners and thieves away, stopping crew from jumping the ship, avoid brawls. The dock police at times proved to be a valuable resource, close at hand. Garberg used his policeman's whistle then, but only when the two mates were at the end of their tether. 'We were armed with revolvers, coshes, and handcuffs' and the two officers made use of them. 'The captain was on land as always and it was the mates that had to bear the brunt'.

Due to the interference of runners, who with frenzy traded crew in British harbours, they were not numerically complete at take-off. Not to upset all registers, changes of the crew was not always noted down in the ship's journal. 'Had one man escaped, then his substitute often

---

⁶ From the number of enrolment registers of these years, complicated by increasing staff and employee turnovers, it transpires that the crew was changed many times and that J. L. Clase was substituted by M. Svan as second mate in Liverpool—who in his turn was replaced on this post by O. F. Westling, who was signed on 19 May 1892. In Garberg's account, Mr Svan appears with his real name without any smokescreen. As will be evident, they were birds of a feather and formed an efficient team in their own right.

took over his name in the book'. To the last minute, there was bargaining going on, and a captain with one man short had to hire another one to exorbitant price in the breakwater: a man without a name.[7]

'Anyway; on 9 February in 1890, the barque was towed through the gate of the Birkenhead dock, even though four men were missing'. The shady transactions were not at an end:

> At the piers so-called pier-head jumpers were ready to board with their bags, being at any captain's beck and call. However, the runners controlled this traffic and secured a half month's salary in advance, otherwise they stopped the jumpers and hailed the captain: 'I have people in a boat waiting on the stream off the gate, so you will get the men needed'. Well, we received the completion we wanted, passed the Holyhead lighthouse, and there in St. George's Channel the pilot boat left us, and we could make sail.

When all work in the rigging was complete and they were under sail on the officer's harsh command,

> the captain expressed his admiration in a flattering way, and then I thought that life was changing for the better. Even concerning the crew I began to entertain hopes. When the alcohol vapours had cleared, there was hardly anyone who refused to work, and a quite good-natured atmosphere arose on our ship [...] Generally speaking, pretty good order and discipline are maintained on Swedish ships, but I had considered it necessary to stand strict on the rules to be able to direct this gang of heterogenous elements.
>
> Now, however, with the captain's consent I slackened the reins as we wanted to keep the pleasant feeling. The master mariner inquired the crew about the fare, whether the food was satisfactory to everyone, and saw to that British manners and customs were

---

[7] Although obviously incomplete, the number of enrolment registers from these years show that crew was indeed changed many times, and that there were lots of foreigners among them.

applied.[8] It might appear that I reigned supreme on the boat, but the captain in fact left most decisions to me, gave just a few orders himself and checked the navigation continuously, but in other respects he relied upon his chief officer.

However, in a little while the harmony got out of tune when the conductor dozed off a bit and each and every one was left to his own doings and judgement in the relaxed laissez-faire policy. Workers before the mast and in lower deck went too slowly and reprimands were met with cheeky replies. Some did not obey orders and were up in arms. The officers then changed their tune. 'Now I had to be equipped with revolver all the time, and the second mate took the same precautionary measure. All instructed work was seemingly apprehended as coercion or punishment and was done aversely and as unhurriedly as possible'.

On a Saturday, when the officer and his larboard watch cleaned deck and gunwale with water for the holiday, there was a quarrel with the forecastle about coveted pots and bowls, things liable to be stolen. At an attempted theft through the skylight 'I instantaneously blaze away two charges in rapid succession [...] whereupon it was dead silent in the forecastle. The scrubbing was finished off without any further incidents'.

Another week, however, the first mate is challenged by the whole watch, which approaches him armed with belaying pins and hand-spikes, but is deterred and held at gunpoint.

---

[8] Concerning diet and eating habits, the British handbook *Cookery for Seamen*, first published in Liverpool (!) in 1894, gives an idea of the current cuisine. Please note the variety of English puddings and cakes, along with cold meat dishes such as Hashed Mutton, Cold Fish Hash, and the ordinary Salt Fish Chowder [*Cookery*, 2019]. Fish & Chips, later a principal dish, was difficult and dangerous to cook in the primitive galley of an old sailing vessel and was only once in the blue moon deep-fried or grilled outdoors at anchorages. Nonetheless, orders for the compulsory Ham & Eggs and Curried Chicken could occasionally be attended to and served—granted that officers, steward and cook managed poultry-house and pigsty with successful care and got subordinates to clean the dung out without stirring up mutiny. It is well known that captains with productive laying-hens on board gained laurels and were remembered with high esteem. They could count themselves lucky. The way to a sailor's heart was through his stomach, and ship's biscuits, soup and salted flesh would not do on the standard menu forever.

3. From Argentina to the United States

'After three month's passage we dropped anchor on the River Plate as the ship was too big for the dock'. The unloading of the unwieldy coal cargo was rendered even more difficult and time consuming by the aggravating circumstance that the harbour offered no assistance by specialized dock workers or bargemen. In addition, the overloaded crew were firmly resolved to jump the ship in Buenos Aires, where runners were at work as in Liverpool. In view of that, a big loyal German among the able seamen is promoted to third mate with pay-increase and moves to the stern castle phalanx as a guardian of law and order. The discharging was dragging on. Thieves boarding in the night are driven away by volleys of revolver shots.[9] The captain has left the ship.

'The master moved ashore and put up at Hotel Söderström. There was indeed a guesthouse with that Swedish-ringing name. Aided by the ship's signal system, we exchanged messages. The captain had access to the international code of signals, and the hotel "Söderström" had the requisite devices that made communication possible'.[10]

To the men struggling on the ship and the lighters, disbandment, furlough, leisure, dockland spree and city amusements were dreams or unknown concepts. Therefore, the breaks for a smoke in the forecastle were prolonged indefinitely. Smoking while you work was not accepted on Swedish ships. Naturally, pretence at working and pretending to be ill were common phenomena.

The real malingerers were cured by an overdose of the purgative *Sal anglicum*, 'English Salt', i.e., magnesium sulphate ($MgSO_4$), dissolved in a glass of water—a prescription medicine 'which completely relieved the victims of sham illness and their inclination to stay in the hammock to smoke pipe and read novels'.

The crew did not get opportunity to do the tango with dark-eyed, black-haired dancing-partners ashore, but were only faced by the

---

[9] His son Sven Garberg, who inherited Rickard's large-bored firearms, meant that they were heavy enough to butcher oxen.
[10] They are using the Code of Signals issued by the British Board of Trade in 1887, which differs from the international signal system of today, which is more detailed and contains a number of modern concepts.

pitch-black coal and soot, as they had been confined to the ship during the stay. Such was their experience of the fair winds of *Señora Santa María del Buen Ayre* in a nutshell.

The only real white-collar worker, the captain, had been busy with paperwork ashore—with the exception of those occasions 'when the signal flags were hoisted and called his attention', and he had to interfere; once in the company of the consul and gendarmes.

Looking on the bright side of things, the chief officer observed that when they 'finally left Buenos Aires in ballast to traverse the Pacific Ocean via Cape Horn, not a single man had disappeared. No runaways, in other words'!

'On our way northwards, we touched at Juan Fernández, the Robinson Crusoe island,[11] for replenishment of perishables [...] a group of islands where people eat langoustes, and of course we also provided ourselves with these delicious crustaceans'.

They were going further north, from South America to North America. They passed San Francisco at a distance and proceeded to the State of Washington in the Pacific Northwest region of the United States. They went past Vancouver Island into the Strait of Juan de Fuca and onwards down south through the Puget Sound basins 'and along their mighty and extensive shorelines, surrounded by a mountain scenery densely wooded by thick-stemmed trees. Finally we reached our destination of order, Port Townsend. The Swedish consul lived there, actually an American, agents and businessmen, most of them timber wholesalers, who obviously lived in a big way from the exportation of the gigantic trees. The so-called Douglas fir, "Oregon pine", can grow to a height of 80 meters and a thickness of almost 3 meters'.

The *Atlantic* was to get its cargo further down the river at Gig Harbor, in the direction of the forest city Tacoma, 'notorious for its wild life'. The region was marked by sawmills, workmen's barracks, and hard forest work. 'The felling was ravaging, and the contingents of lumberjacks had

---

[11] The allegedly uninhabited Chilean island which is said to have been the home of the marooned sailor Alexander Selkirk in the beginning of the eighteenth century—whose story probably inspired Daniel Defoe's myth-making novel, which in its turn propelled a whole genre of literary, philosophical and environmentalist *Robinsonades* and the modern ideology of self-subsistent households and back-to-nature lifestyle.

to move from place to place to supply the demand'. Working hours were from five in the morning to eight in the evening. Breakfast was served at half past four and dinner at eleven thirty. In addition to three meals a day a boy could earn five dollars, and adult men ten to fifteen dollars a day. 'In these circumstances it was impossible for our captain to keep his crew'. His hold on them was four months of withheld payment, which they would lose, but before too long 'the bow door was open, and bags and men disappeared from the forecastle to the waiting runner boat until there was only the afterdeck guard left'.

This meant that even the galley was deserted, and that there was no more cooking on board. Yet on land the dining hall of the sawmill workers was open to those who paid their way, and was irresistible to all starving men: 'offering delicious roasts, ham, pies and puddings, plenty of eggs, milk and cream, coffee, tea, and a choice of pastry; an abundance a first class hotel could not excel'.

The captain and his decimated crew focused on other tangible assets. 'The recently built sawmill which supplied the ship with wood products was equipped with a circular saw. Each log was sawed up in thick planks which were put in a timber yard, exactly as in Sweden. Most freights were to Australia and China. Besides plank we took in huge beams dimensioned 36" × 36" in lengths of 110 feet. These whoppers reached to the bowsprit and required heavy lashings'.

Injuries and deaths were not uncommon among the workers, and one day the cocky and recalcitrant officers of the *Atlantic* gained popularity on land by participating in a funeral, at which 'the unusually large ship's bell' serves as church bell in the ceremony. This was widely acclaimed, and the assembled people came to the ship afterwards showing a sympathetic attitude. 'Even the captain, who had kept in the background during the stay on shore, came forward to be honoured by the workmen'.

After bargaining with the wholesalers, negotiations with employment agents and tough landlubbers followed, but finally the berth was full, the crew complete and the ship ready to sail. In reality, going to sea was attractive to many on land; they had been waiting for this opportunity. The captain, who was an experienced cashier and a sly fox, knew the lie of the land and held his own in the haggling, and knew how to get the best of it in the end. Remember that the outstanding accounts of the lost crew had been nullified.

4. Steering for Australia and the Cape Horn

Thus they set off 'with a glorious mess of people', bound for Australia. There is scuffle on board, whereupon the captain places two troublemakers under arrest for the rest of the trip, which restored order. However,

> when we had left Melbourne to load sacks of wheat at Port Germein in the Spencer Gulf, the forecastle was soon empty again as the high wages of reapers during the harvest time draw all crew from the ship. Hence new men had to be hired, and this time they were less skilled—cowboys, sheep-shearers and men from the bush, which had no idea of life on the sea, rigging and sails. All had taken on under the pretext of being able seamen, and demanded to be well paid for their work too—six pounds per month, that is, about 108 kronor, whereas I had a monthly salary of 75 kronor. And now I was put to set all things right and train these people to become sailors.

There were other challenges, too, the weather and the chosen course. They were homeward bound, first heading towards England. In March 1891, the cold-blooded captain had fought many days against the stubborn Westwind in his attempts to set out upon the so-called summer route and go westwards back. This proved too difficult, however, and he finally decided to veer and go eastwards instead, in the opposite direction, relying more upon the fair wind than the aptitude of his new crew. This also meant a more southerly course, to deviate a bit from the fortieth parallel of the Australian south coast and going down toward the southernmost headland of Chile at 56° South, considerably closer to the Antarctic region.

It was chilly on board, and the cold was on the increase, it was biting and went right through the bodies. The crew began to murmur and grumble. 'I felt really sorry for those poor creatures who had no clothes. Some of them fared so badly that they could not be on watch, the others had to do hard work then. Usually I had four men on my watch. A sailor certainly understands what it meant to manage a 1,000 ton ship rounding the Cape Horn in the coldest season with this defective help'.

The captain at the steering wheel, who was also the treasurer and stores manager, knew how to derive advantage from the plight. When the crew approached him and asked whether there were not any cast-offs in the refuse or the store chest that they could buy, the business began. 'He did have some, but wanted to get a good price for it... "Be freezing or pay" became the captain's watchword regarding these problem children of Neptune'. The overpaid crew had to pay back. Three pounds for Wellingtons, five pounds for socks, all was sold out at exorbitant prices–knitwear, jerseys, oilskin clothes, shirts, trousers— and everything was put down to the account of the person concerned, a sum to be knocked down from the salary at sign-off. All these deductions were finally checked by the Swedish consul, who also supervised deduction from wages due to incompetence or bad behaviour documented in the ship's journal.

The chief officer for his part manufactured wading boots with legs from sailcloth and filled them with straw. It was cold, indeed, and wet. Cheer up, the worst is yet to come!

There was a strong wind blowing too, and it was rising. A storm was brewing, and before long they were hit with hurricane force. They take in sail, wait, see nothing in the storm centre. 'We only hear the roar, the waves running high over the forebody and come dashing in, over and over again. Nobody dares to move on deck. The crew is secured, lifelines are stretched over quarterdeck and cabin roof, to hold on to when needed. Two men manage the wheel, both tightly tied up. The captain is in the stateroom. Myself I am standing on quarterdeck at a hood with a strop round the waist, close to the lifeline'.

The seas beat with terrible force against the hull and splash over the gunwales—and then suddenly it is 'as if the ship is pressed down into the ocean, drowned from stem to stern in a crashing and suffocating breaker. Then pretty well every one of us believed that we had breathed our last', but the ship surfaced and reappeared. Catching breath in the sudden lull that followed, and getting over the shock, they saw that the ship was not the same as before. All was different.

'It was as if the sea recovered its breath after the heavy blow it had administered, and we all came to and could look around. The hood to which I had been lashed down was gone, but strangely enough I was still there, although badly knocked about. I was bleeding from ears, mouth, and nose, and I was hanging in the lifeline halfway over the wale. I had been unconscious for a while. The man on the opposite side of the hood had been caught by the torrent and was engulfed by the sea.

Thick iron bars ahead had become crooked; the two lifeboats on the forecastle roof were damaged. The railings on quarterdeck and cabin roof had disappeared. The steering wheel was in pieces; binnacle and compass gone; the two helmsmen lay tossed up on the mizzen boom, still fastened to their ropes. The rest of the crew were standing in their lashings sheltered at the mizzen rigging. Both watches had been on deck and had they not all been made fast, the captain had soon enough been alone on his ship'.

Besides that one man was lost, the material damage was extensive. The port side gunwale was demolished, the wind turbine destroyed, the main hatch crashed—which was not discovered until the wind was abating at the light of early dawn. Then they realized that the tarpaulins had stopped the streams from penetrating the ship and sinking it. The cargo hatches had been covered by 4-inch plank with strong crossbars and was heavily lashed to eight ring bolts on deck, and all this was broken and the bolts extracted from the wooden vessel by the forces of nature.

'The captain we had almost forgotten in the confusion. He was swimming in the saloon, which had been half waterlogged after the skylight on the cabin roof was dashed to pieces. Luckily enough, the large square sails were down there, which the captain planned to darn and stitch up, since there was now no one else on board who could do sailmaker repairs'.

The weather was clearing up, and when the ship was pumped dry, the hatches roofed and the cabin covered by a sail, securely nailed, the trip continued. As the sun rose at the Cape, the starving sailors took an interest in petrels and albatrosses and became bird-catchers, and when they were hit by the favourable south-east trade wind between South America and Africa they turned to fishermen with almost identical tackle, supplemented by jigs, spears, and nets, and the crew out of the blue excelled in their skills.

When all these men were signed off in Rouen in France, some of them made a terrible scene when the disbursement took place, 'but the consul was inexorable and insisted that all should be paid according to the book. One of those who remonstrated against these calculations was very aggressive and refused to accept a penny until he had talked to the king of Sweden'!

With a completely new crew—once more!—they returned to the kingdom of that sovereign, who only talked to mariners if they were lieutenant commanders in the navy.

In his account, the chief officer's dreams and thoughts had lingered in the memories of all the nice uses of the albatross, when he considered that he had done his last Horn passage on this route—how the magnificent, white gliders of the southern oceans were hooked by trolling with a long line of a hundred fathoms, and having swallowed the bait 'was landed on the taffrail in the stern and beaten to death with a wooden belaying pin'.

Mariners, ancient and modern, made use of the entire albatross. Breast and thighs were chopped with salt meat and made into meatballs; the down went into pillows and quilts; from the bills hangers and inkpots were made, and wing-quills became pipe shanks. The giant petrels were more appreciated as a delicacy, cooked with the commonly stored condensed milk, butter, and spices; considered rich and scrumptious food and a touch of class. It is true that the sea air gives you an appetite. This is the way of the world.

5. Life on Land

The first mate Rickard Garberg was happy to be back with his wife on 14 September 1891 after having earned 1,627 kronor and 50 öre on this 21-months' trip, most of the money sent home during the journey. His matrimony bore fruit, but true to his wife he was also true to the cash cow *Atlantic*, and when Axel Söderström called for his assistance again he could not refuse it. They set off in May 1892, to be away for three years! The master appreciated his well-judged and resolute factotum who did the rough work and kept order. Owing to this, harmony was maintained in the stern castle. The navigator only asked for peace and quiet so that he could bring the ship to harbour and his business matters to a successful close.

Captain Söderström, on the other hand, right now looked forward to attending the wedding ceremony of his daughter Emmy in Gefle! She had returned to Sweden beforehand in company with her crestfallen cousin Leonard, the dejected second mate, who would participate in the ceremony, although it was somewhat awkward and painful. Relations between the aunts was already strained. Nevertheless, a peace dove always hovered over these nuptials and their ensuing friendly talks and celebrations, which were reconciliating to so many families.

On 29 September 1891, the 23-year-old globetrotter Emmy, by all close friends known as 'Emy', espoused the harbour official Gotthard Hallengren, 34 years. As a cashier and clerk he supervised the shipping of Sandvik Steel. As it happened, Emmy married the son of one of those who had organized the rescue operations in the conflagration of 1869—the disaster where Emmy and her sister had survived outdoors.

The spouse's sister Anna, married to a consul, was the mother of one of the future owners of the shipping company PJ Haegerstrand, which was one of her husband's employers—the company where two of their children, the daughter Greta (b. 1892) and the son Bo (b. 1900), would be agents and owners. Another son, Stig (b. 1903), became travelling salesman for Sandvik Steel. The family remained deeply involved in international trade and shipping, always playing an active part in the business.

The tempo and the way of life was brought by the sailing ship and the widened horizons of seafarers. Yet the *Atlantic* now had to overwinter on land in the shipyard and her captain devoted his time to family life and future prospects. He had become unfamiliar with town life, and people at home were unaccustomed to see him. He stepped into the cobblestone alleys anxious about his boat in the dockland and sprained his ankle. The ship had become an object for renovation.

The damages caused by the hurricane at Cape Horn also included considerable loss of equipment and interior furnishing and fixtures, and there were many other minor details to look after.

So much had to be replaced: banisters and railings; companion way; spiral staircases; quarterdeck stairs of oak with metal mountings; quarterdeck storage space; cabin mouldings and frames in addition to new sliding doors; three new cargo hatches; after hatchway hood; a new pigsty (nota bene!); boat davits; lifeboat with equipment

(450 kr); gangway ladder; 4½ feet steering wheel (125 kr); binnacle; meat repository of mahogany; spirit compass; lifebuoy; telescope; log glass; light; cabin furniture; rudder post; scupper, etc, etc. In the exterior, and on the larger scale, there was planking, caulking, renewed coppering of the hull to be done—carefully avoiding all metal since copper and iron react galvanically in the water, causing corrosion—; sails and masts to mend, and everything was listed and put on bill by the Brynäs Wharf.

During this time the winter of 1891–1892 passed. Axel's newly wedded daughter was pregnant—but would the maternal grandfather attend the baby girl's baptismal rite in July, slightly more than nine months after the wedding?

Being independent and his own master, yet a married man longing for a family, the seaman Rickard Garberg was on the lookout for work on a steamer in the spring, trying to avoid the yearlong journeys. This was unsuccessful, and when the captain of the restored *Atlantic,* which looked brand new despite her age, offered him employment again, he could not refuse it. Still, it cut him to the heart—exactly as last time, when he hesitated to embark in Liverpool.

6. Time Lapse: Long Voyages in a Changing World

No sooner said than done. The captain missed the baptismal font and the christening with water, and the pregnancy of Mrs Garberg had apparently escaped the first mate's notice in the haste. On 17 May 1892, all crew had been taken on, and the ship in Gefle harbour was ready to start. They loaded 2,021 cubic metres of wood products for the forest industry company Korsnäs AB, and departed for Argentina. PJ Haegerstrand was the new shipping agent, the employer of Axel's new son-in-law.

They arrived at Buenos Aires in September. Chief officer Garberg remembered: 'Then a wire arrived, informing me that I had got a son. I was overjoyed. My one thought then was to save money, to practise a strict economy, and take a bright view of the future'!

Time flew, while the ship made progress at an amble. They had errands and business journeys to various European ports, and then they were to the Port of Pensacola in Florida in the Mexican Gulf, an American possession since 1821, in earlier times Spanish. This trip was as adventurous as the former one, although the captain was not swimming in a flooded saloon this time, and it lasted even longer. Finally they loaded 1,517 tons of coal at Newcastle in England for sea transportation to the Gefle Gasworks, and the transactions were still administered by the shipping agency PJ Haegerstrand.

They arrived at Gefle on 2 July 1895, after over three years absence from the homeport, arriving old as the hills from the past in the days of sailing ships. The chief officer was jubilant.

> 'Time was getting on and finally we could write 1895. I came home then, and I never forget that moment when I returned from my last long journey, full of good intentions and hopes, since I had a wife and a child to take care of and to be responsible for; and how I knocked at the door, and a three-year-old towhead opened the door slightly, and then cried into the room:
>
> —*Mummy, it is a man out here!*'
>
> [Garberg, 1962, 97–128]

To the captain there were also surprises. Emy suddenly has two sons also, and her sister Mia in 1893 had given birth to a baby girl, named after the ship *Gurli* where her father John Ljungberg was still the captain—and where she was herself occasionally staying. According to a family legend, a bird of prey one day swooped down upon the baby on deck. If this ever happened, it was probably just an incident; shocking, to be sure, and for this reason remembered. The sailing ships were visited by birds of all kinds, snatching up what they could find of scraps, offal, fish and chicken. Nonetheless, the story at least indicates that John from time to time was still sailing with his family on board.

7. Söderström's Long Farewell Voyage

After having seen Emmy's family at Byggmästargatan 9 in the Third Quarter—with three new family members born in less than three years—the grandfather returned to his element and put to sea in August 1895 in his windjammer. It was the year when Joshua Slocum set sail for his single-handed circumnavigation with his sloop the *Spray*, a three-year voyage for which he had been trained by navigating the barque *Aquidneck*, and during which trip the lonely sailor encountered a profusion of large sailing ships on the way—barques, clippers, full-rigged ships—and next year the traveller Rudyard Kipling, born in India, issued his poetry collection *The Seven Seas*.

Axel Söderström is on a new long roundtrip navigating his merchantman on the oceans, shuttling between producers and purchasers worldwide. Signing on in Gefle on 16 August: embarkment and departure on the 24th for the sawmill in Svartvik, Sundsvall, going in ballast to fill the hold with timber. He signed off in Hudiksvall on 15 June 1897 after having carried his commissions to a successful conclusion after hardships, nasty weather, and terrible accidents. The Swedish daily *Vestkusten*,[12] published in San Francisco and Mill Valley, California 1887–1991, carried the following news in Issue 21 on 27 May 1897:

> A young resident of Söderhamn, the deckhand [jungman] Axel Trapp, employed on the ship Atlantic of Gefle—navigated by Captain J. A. Söderström—died in a fatal incident during the ship's voyage from Sabine City in Texas[13] to Belfast in Ireland on 2 April. The accident took place during a terrible hurricane, which hit the ship near the Azores and the violence of which was unmatched according to Captain Söderström, who said that he had not witnessed anything equal. During the battle against the raging storm no one noticed how the accident happened. It is probable that Trapp, together with a comrade who also perished, was thrown overboard by the high seas that washed over the ship, which also carried a deck cargo of wood products. Axel Trapp,[14] who now made his first long trip—he enlisted on the Atlantic in September

---

[12] ISSN 1073-6883.

[13] The town of Sabine Pass, Texas, originally chartered Sabine City, was established in 1839; it was projected to be a major Gulf port with cattle and cotton the port's major shipments. Eventually the name was formally changed to the more commonly used name of Sabine Pass in 1861. Hurricane and storm damage led to the decline of the city, which is now listed among the ghost towns.

[14] Axel Trapp was born in 1878 in Söderhamn, Sweden, became a member of the crew on 11 September 1895, with a monthly salary of 15 kronor, and is reported to

1895—was a promising young man, who seemed to have a dazzling future before him. He was 18 years old and the son of harbour captain Ernst Trapp in Söderhamn.

The captain, never to be intimidated and resolute as an engine, went on in a couple of weeks, signed a new team on 8 July in Hudiksvall and went to Cape Town, South Africa. PJ Haegerstrand was arranging this new long trip, which did not end until 1899 when they returned to the homeport on 26 May from Wismar, the German and former Hanseatic seaport, loaded with fertilizer. Captain and crew disembarked the next day. Having emptied the hold and replaced the cargo with ballast, on 18 July they went up to Sundsvall, where they loaded timber in Kubikenborg next day. It was a shorter trip this time to be continued after summer (Figure 29).

The captain became jaded and began to fall away after twenty-three years as master mariner of the *Atlantic* and forty-four years on the sea. His motherless daughters fretted about their father. The sisters still mourned their dear 'Grandma', the foster mother. Maria was furthermore concerned about her frail and vulnerable little Dolly Barbro, born on 20 June 1899, whom she nursed in Västervik, staying there alone with her children, missing her husband who was far away on the sea. A dramatic change in the situation was now in the offing, and they all shared a presentiment of hope mixed with fear.

8. Murder Mystery in Cape Town

Axel Söderström, who never remarried, moved to Emmy's new home in residence No. 175 in the Third Quarter, which according to the Census of 1900 houses two families, one of which is a widower, living alone. The other family consists of his daughter's family of six persons, including husband, and now four children.

The aging mariner's mind is made up. He has come to a crucial decision. In agreement with the shipowner P. A. Engström, he offers his post as captain on the *Atlantic* to his daughter Maria's husband

---

have signed off dead on 2 April 1897 [Sjömanshusets i Söderhamn arkiv, the Swedish National Archives].

John Ljungberg, the master of the *Gurli*. The captain is succeeded by his son-in-law.

The crew of the *Atlantic* is signed on in Härnösand on 7 September 1900, with Captain Söderström at the helm. They were all busy taking in a huge cargo of planks there. Heavy and tiresome to be sure, but nineteenth-century sailing ships sometimes had to accommodate elephants, which made for a lighter, although pretty more unwieldy cargo than dead trees. Carrying responsibility, too. Consider the air of young deckhands and constables on watch in the seaway!

On 27 September 1900, the captain signs off. Next day John Ljungberg, 48, ceremonially sign on as the captain of the ship. His wife Maria embarks the ship with their children Axel, Gurli, and Barbro to be present at this impressive occasion. Now she is back to the *Atlantic*, once more the centre of her life. The atmosphere is solemn. They are all experienced seafarers. First mate and chief officer of the ship since 1899 is Axel Söderström's unmarried nephew Knut Reinhold Zellinger, his sister Charlotta's son, 38, who had admired his cousin Maria since his youth when they lived in the same homestead.

All's well that ends well? Maybe. Yet this family constellation cut both ways. In this harmony, nobody was prepared for the worst. It was the beginning of the end.

The worst happened! On 21 March 1901, captain John Ljungberg was killed in Cape Town, South Africa. According to the chief officer, he died when they went on an airing together in the seaport. There were rumours that he had been beaten, had drowned in the harbour, and other guesses. The shock cut so deep into the hearts of the family that the aftershocks and repercussions descended through the generations to the present day. Speculations about murder, a *crime passionnel*—that the captain was murdered by his rival, the first mate, who took over the ship as its master and navigated it henceforward [Herrström, 1973; Henricson, 2003; Kjellgren, 2014, 2020].

There might have been some competition and discord between the two, but a mountain was made out of a molehill. There are no contemporary indications of committed crime. The widow apparently never thought so, although the *pro tempore* captain later wooed her and is said to have written her an indignant letter before he died in tragic circumstances two years later. The particulars of the case are as follows.

The *Westervik Dödbok* (Book of Deaths and Funerals) in 1901 notes that sea captain John Erik Ljungberg passed away in Cape Town on 21 March 1901 referring to a telegram from the consulate-general in Cape Town of 22 March, and a writ from the main shipowner of the *Atlantic* P. Engström in Gefle dated the same day. *Slag* ('stroke') is stated as the cause of death. Taken alone, the diagnosis is certainly ambiguous, in the first place as it could either refer to an ischemic or a haemorrhagic attack. These days, the term most often indicated apoplexy, a bleeding in the brain, but diagnostics was far from a refined medical practise. Certificates of death were often arbitrary.

Out reporting in South Africa in the 1990s, this author tried to find further substantiation of the case, but due to the reorganization of government and constitution, this proved difficult. A grandchild of Maria Söderström's daughter Gurli, Thorleif Herrström, was a cleverer researcher and ten years ago successfully acquired copies of the available South African documents of 1901.

The first entry is from the shipping register held at the Western Cape Archives: Report of arrivals and departures from Table Bay in 1901. There it is recorded that the Swedish barque *Atlantic* of Gefle with a tonnage of 1,032, carrying timber cargo, J. Ljungberg the master, arrived from Härnösand on 5.05 p.m. 1 January, the date of sailing given: 3 October [1900; which refers to the final departure from a Swedish harbour]. 'Date of departure from Table Bay 11 April [1901] at 10 a.m'.

The second document is the Death Certificate, No. 520/1, concerning the master mariner of the barque *Atlantic*, a married man of 48 years and European race. 'Date of death: 21 March, 1901. Place of death: Somerset Hospital. Intended place of burial: Maitland Cemetery. Cause of death: Nephritis, Uraemia, Coma. Duration of disease or last illness: 3 hours at hospital. Name of medical practitioner: E P Collins. Informant: J Veitch, Somerset Hospital. Date registered: 22 March 1901. Place registered: Cape Town. HAWC 1/3/9/5/5' (Figure 26).

Consequently, captain John Ljungberg died from a kidney disease, an inflammation. Renal failure fell upon many sea captains, and nephritis is still a very common cause of deaths worldwide. As is known from the Westervik Book of Deaths and Funerals, John's father captain Sven Reinhold Ljungberg also died from an inflammatory kidney disease,

DEATH CERTIFICATE

Entry number	520/01
Christian name(s)	Jan E
Surname	Ljunberg
Sex	Male
Age	48 years
Race	European
Birthplace	Sweden
Marital status	Married
Occupation	Master mariner
Pensioner or dependant of pensioner	-
Date of death	21 March 1901
Place of death	Somerset Hospital
Usual place of residence	Norwegian barque 'Atlantic'
Intended place of burial	Maitland Cemetery
Cause of death	Nephritis Uraemia Coma
Duration of disease or last illness	3 hours at hospital
Name of medical practitioner	E P Collins
Informant	J Veitch Somerset Hospital
Date registered	22 March 1901
Place registered	Cape Town
Notes	In HAWC 1/3/9/5/5

Certified as a true copy of the original register held at the Western Cape Archives.

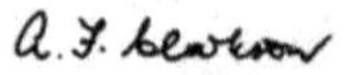

A F Clarkson
A F CLARKSON

Fig. 26. *John Ljungberg's death certificate of Cape Town 22 March 1901.* Private possession.

Morbus brightii, on 22 January 1868. *Uraemia* means a raised level in the blood of urea and other nitrogenous waste compounds that are normally eliminated by the kidneys, causing intoxication. And *coma* is a prolonged state of deep unconsciousness.

Poor captain! Poor officer! Poor families! All suffered badly and unjustly. In 1903, Knut Reinhold Zellinger's mother, the widow Charlotta, received the news from Chicago that she had lost her second son too, reportedly by suicide, and it is told that he hanged himself. The ill-fated Aunt Charlotta Zellinger was reduced to despair—again. But we know that there was no murder, and that the chief officer was as innocent as unlucky. Yet, the note about the 'stroke' remains ambiguous and mystifying.

When Knut Zellinger has been appointed acting master mariner, as confirmed by the consul-general of the United Kingdoms of Sweden and Norway in Cape Town on 8 April, 1901, they put to sea in ballast and sail for Pensacola in the Florida Panhandle in the Gulf of Mexico. With a shipload of pine wood they leave for Tunis. Captain Zellinger is skilfully steering the large sailing ship forward, carefully navigating it through the Scylla and Charybdis into the Mediterranean without pilot or tow. After Tunis they continue with remaining cargo to Bizerte on the northern coast of Tunisia, landing on 6 November; and then they are to Almería in Andalucía, Spain, on 5 December 1901. The complex and slow freight route across the seas can be followed in consular reports in a number of cities. From Almería they go far westwards again, and set sail for Santos on the coast of Brazil, and on 26 March 1902 they are cleared outwards for Barbados in ballast. Everything goes on with the same procedures as always, zigzagging between loading ports and places of discharge. Next: the port city Kingston upon Hull in East Riding of Yorkshire, one of the *Atlantic*'s regular destinations, and finally departure for the homeport on 13 September 1902! On 27 September 1902, they are back to Gefle, arriving from Hull with heavy coal cargo and a draught of 21 feet. Captain Zellinger had then already been replaced by captain Otto Olsson, Gefle, who served as master mariner 1902–1904. His predecessor went to America, while the *Atlantic* spent the winter of 1902/1903 in the Gefle harbour.

9. The Last Days of the Skipper King

Maria's husband John was buried in consecrated earth, although not in Sweden but in the Western Cape province. The black memorial stone of diabase—now a listed monument of ancient culture, which his widow subsequently raised in the old Westervik Graveyard—for this reason does not carry his name. Instead, it carries the name of her father, the ancient mariner Axel Söderström, who passed away the year after her spouse, in 1902, and is buried there (Figure 27).

He had moved to the house in Västervik to support his widowed daughter, now a single parent with three children.

Some days during that last year, they had a really good time together. The old salt enjoyed being with the children and bringing up pleasant memories with his daughter, and the original grandpa enlivened the atmosphere with the old gear he had brought in the move. His well-lined wallet ensured that nothing was missing in the house, and Maria received a considerable legacy from her husband's estate. They allowed themselves everything they wanted. People in town remembered the old man, dressed as a Turk with fez on his bald head—the flat red hat with black tassel on top. At home he wore a Turkish smoking cap with smoking jacket and enjoyed to the full the cool fragrant fumes of his bubbling hookah, filled with strong oriental ship's shag tobacco from his pouch, sitting in a suite of oriental furniture—a few exceptional items still preserved in addition to a pipe bowl.

Arabic coffee with sugar, syrupy punch and Moroccan sweets were devoured, aggravated his diabetes and in the event seriously exacerbated his kidney problems. Life in the tropics with infections, malnutrition, beverages of variable quality, humid cuddies with no sanitary articles, in conjunction with shift work, hardships and perils, shortened the life of many middle-class master mariners and not only the lives of common seamen. As we have seen from his daughter's travel journal, he was hospitalized in Australia in 1887 for typhoid fever.

Bedrooms in the houses on land were always upstairs because it is usually warmer there, and before too long he spent more time in his chamber, resting in bed. Nearing the end, relatives remembered his forced breathing, his horrible panting. Finally, he was brought down to the ground floor. These days, it was common that the dying lay in a wooden sofa bed close to the entrance, often with a lid that could be closed when all was over. Carpenters made their own coffins and lay there the last days in waiting for the day of doom.

Axel died in Småland, where his grandfather, the lieutenant commander Jonas Söderström had been born in 1774, so in a way he returned to his origin.

On 23 February 1902, a funeral procession passes over the Old Cemetery in Westervik. The one to be buried is sea captain Johan Axel Söderström who at the age of 61 died living with his daughter Maria and her children. She had about a year earlier lost her husband, sea captain John Ljungberg, in Cape Town. The funeral party also includes their children, Gurli, Axel and Barbro. Maria later has the grand tombstone erected in memory of her father and two of her children—the daughter Dolly Barbro Ljungberg tragically passed away next year, 4 years old. [Westervik Grave number 1-0401.] Maria had lost parents, husband, and her youngest child. At 26, it remained for her to take care of the living and look forward. In 1905, she remarried, but never regained the happiness and excitement of her youth, solicitous about the welfare of her dear children but wandering between different Swedish addresses, moving in a cul-de-sac, as it were.

After the funeral there was the estate of the dead to be taken care of—the trying practical and legal obligation incumbent upon all mourners, upsetting to all surviving relatives, as private and intimate possessions and all worldly goods then are reduced to material things with price tags and put in boxes. No wonder that items with but sentimental value are those preserved. A pipe bowl. An old sacking easy chair. Such things. They have lasting value.

The Westervik Book of Deaths and Funerals of 1902 registers the death of sea captain Johan Axel Söderström on 18 February. Cause of death: '1. Nephritis chronica 2. Uraemia'. Compare this to the death certificate of his son-in-law! Both suffered from an inflammation of the kidneys which proved fatal.

On 26 February 1902, the estate inventory proceedings took place in the presence of the two heirs, the widow Maria Mathilda Ljungberg and Emmy Axelina, authorized by her husband the cashier Gotthard Hallengren, absent. In the deed there is the usual pocket watch in gold which all men of the middle class owned along with a wall clock; in addition, a divan, a desk with swivel chair, various books, an iron

Fig. 27. *The family memorial in Westervik. Maria Söderström raised the stone to the memory of beloved father, daughter and son; but her spouse does not rest there.*

bedstead, common furniture, two Emma chairs (upholstered easy chairs), household utensils, a washstand, barometer, wearing apparel, etc. The listing is a remarkably short selection.

Instead, the focus is on the remarkably long list of deposit accounts, i.e., bank balances; interest claims remaining to be drawn, and a life insurance policy. The grand total is 33,340 kronor and 98 öre. The estate personalty added, the full sum amounts to 33,728 Swedish kronor and 98 öre.

Comparing to upper-class luxurious commodity, you could then buy three brand new Mercedes DMG automobiles and a muskrat coat for that money and get an annual salary in return as pocket money. Or, perhaps an old barque as the *Atlantic* or one or two deregistered slave vessels on the second-hand market of ships. By a more modest and realistic, as well as saddening comparison, the commencing annual salary for a governmental employee or a research assistant at Statistics Sweden was 1,500 kr, while steel industry workers with a sixty-one-hour working week earned 0.36 kr per hour. Female workers in the tobacco industry with a fifty-eight-hour working week had hourly wages of 11 öre (0.11 kr) [Lagerqvist, 2011].

The orphans Mia and Emy came into a little money.[15] When they finally pocketed their inheritance, they were both millionaires by modern standards. They had even more purchasing power than the depreciation of the value of money can tell if the total balance were converted to two hundred thousand euro or so. Manpower and service were relatively much cheaper than today, as were most everyday commodities before the socialist equality reforms later in the same century. The class society, which had replaced the four old Estates of the Realm—nobility, clergy, burghers, peasantry, and the nameless masses with no say or pay—culminated in prodigious income gaps in the modern industrialized society in the era of continuous urbanization and the depopulation of the countryside. Emigration had already peaked. People fled the country, running for their lives. A fifth of the Swedish population (over one million) emigrated to the United States of America between 1850 and 1910, a minor amount (over 30,000) to Australia.

10. How Does This End?

Emmy's oldest children Greta and Sven (b. 1892 and 1895, respectively) talked about the marvellous moment at the turn of the century when the oil lamp was replaced by electric light. In the street, a carriage suddenly rolled forward without a horse—thereafter called *automobile*,

---

[15] Swedish women were by law entitled to inheritance since 1845.

'a self-propelled vehicle', and its engine measured in horsepower. In a small funnel, one day they asked for number 99, and the whole family, reverently gathered, heard their father's voice 3 km away—from the telephone at his port office. Another day a little later he brought home a banana, an unknown fruit to the young, which was carefully cut in pieces and fairly distributed among the seven children—an outcome of modern refrigerated sea transportation between the continents.

That was just the beginning. They happened to be born in the time of change, in the dawn of the modern age. No one in the history of humankind has experienced such profound changes during his or her lifetime. It was an accelerated movement through time with unrelenting technological force. They were all convinced that this speed would increase indefinitely—but all agreed that the washing machine was the unmatched invention, remembering women's drudgery in the past.

In 1904, the geographer and explorer Otto Nordenskjöld returns to Sweden from his research expedition with the ship *Antarctic*, 1901–1904. The map of the world is altered by the cartographer who drew a partially new one and introduced new place names.

The *Atlantic*, which set off again in 1903 with its new intrepid captain Otto Olsson as helmsman, continues its roundtrips as merchantman in international traffic carrying on the export and import of timber. On a voyage from Gulfport, Mississippi, with a cargo of pitch pine, the *Atlantic* is damaged in a hurricane in the North Atlantic some 900 miles west of Portugal and puts into the harbour in the Azores as a wreck.

At this stage, the Azores are well known to the reader (Figure 20). When Captain Axel Söderström was struck by the hurricanes of 1874 and 1897 he lost crew and craft here. His *Thor* had foundered in this zone, and all the crew left the sinking ship in the high sea.

At the captain's protest and in the inspection by the Portuguese maritime authority, the vessel was found unseaworthy, and was consequently condemned. It was officially declared unfit for use and sequestered. The insurance company would not pay.

After much deliberation and bargaining, the master mariner Otto Olsson finally managed to redeem the *Atlantic*. On 13 June, he bought the ship on auction for 5,200 *milréis* and got the ship repaired.

The currency used in this transaction was the monetary unit of Portugal, equal to one thousand reals. The very word *mil-réis* literally means one thousand (mille) *réis* (reals) and was an outcome of the nineteenth-century economic crises with accompanying depreciation and devaluation of the currency. In 1911, the *milréis* was replaced by the *escudo*.

The price did not prove prohibitive. The shipbuilder O. A. Brodin purchased four-fifths of the shares for 12,000 kr on 20 June 1904 and took over as shipowner. On 22 June, the *Atlantic* finally returned with ballast from São Miguel Island in the Portuguese archipelago of the Azores and stayed the winter in Gefle—the crew getting off with no more than a fright. Time for a winter resting period and to lick one's wounds.

On 29 June 1905, they are on outward passage from Gefle, going ballasted to the old sawmill community in Skönvik, Norrland. On 15 July, a new master mariner is appointed and immediately takes on, Janne Troedson Nordgren from Örkelljunga, with a degree in nautical science from the Navigation School, class of 1883. He signs off in Örnsköldsvik on 11 November 1905, rejoining the crew the very same day.

Then they struck the Swedish–Norwegian Union flag and henceforth sailed under the Swedish colours, as the union was breached in the autumn of 1905. In 1905 they carry out two journeys in The North Sea and the Baltic Sea. The ship keeps on the surface of the water all year, after her loving care at the wharf in the winter of 1904 to 1905.

Next, the *Atlantic* turns into a virtual Transatlantic Liner and makes four voyages to America each year in 1906–1907; six in 1908. The number of America Swedes is growing along with the traffic and the trade; passengers and goods have become compatible sources of income. On 14 November 1908, the ship is back at Gefle; there is a note in the ballast journal on 20 November. After that the *Atlantic* stays in the harbour during the winter and all of 1909, and the next winter too, resting as before in safe winter storage at Brodin's Wharf by the Gavle River.

On 29 March 1910, a new bold captain casts the die, Oskar Gustav Nilsson, Vitemölla, with an officers' diploma of 1905. Loaded with 1,962 cu. m. of wood products from Korsnäs they depart for London on 18 April 1910, and during this year they carry out six voyages in The North Sea and the Baltic (Figure 28). On 19 October 1910, they return ballasted to Gefle from Stockholm, thenceforth eschewing the sea in winter as before, avoiding the ice—the reason why captain Söderström

Fig. 28. *The ATLANTIC in full sail, 'Captain O. G. Nilsson,' 1910.* Public domain. Courtesy of the National Maritime Museum, Stockholm. Fo80338A.

Fig. 29. *The Gavle River harbour with the ATLANTIC and other ships in 1899.* Courtesy of the National Maritime Museum. Fo24989AB.

always spurned the home port! Captain Oskar Gustaf Nilsson will go on for another year, beginning to find his feet.

On 31 January 1911, the ship owner O. A. Brodin dies, but his company survives with his son Erik as executive. This shipping company will later split into a number of subsidiary companies with new names and the group changes hands, although the original Brodin Shipping still exists and travels the oceans with a number of ships.

In the early twentieth century, however, the number of merchant sailing ships in international traffic were diminishing towards non-existence, and so were the number of shipping offices in Gefle. At the turn of the century, there were six shipping companies with no more than sixteen sailing ships operating according to the Ship List of 1900. The famous Gefle fleet of sailing ships, which in 1842 constituted the largest merchant navy in Sweden, was shrinking rapidly; the illustrious shipyards and the shipbuilders were disappearing from the scene. Today they are all gone, the old harbour barred by bridges, and the seaport almost forgotten.

The barque *Atlantic*, which had been running in the tramp trade since 1876, was taking wind on its last gasp in the swan song of 1911, when she does her last six voyages in The North Sea and the Baltic Sea that year. It was one of the last in a row of over a thousand ships.

To get a true view of what was really happening in 1910–1911, we will examine the last logbook minutely and pick up a few details of interest. Ship's journals were by law kept on all boats in international trade and should be stored by the shipowner or the shipping office for two years. Due to the number of these handwritten books in the merchant fleet, the prescribed filing time was strictly limited.[16] Consequently, deposit copies are rare finds. As this original item in our case is extant, it deserves a closer study.

The last ship's log of the *Atlantic* was kept day by day by Captain O. G. Nilsson from 1 April 1910. The book, prepared and ruled for this purpose, was issued by *Gefle Sjömanshus*, the mercantile marine shipping office to which captain and ship belonged. It was signed by the ombudsman Göran Moberg—who succeeded Johan Axel Söderström as captain of the *Gustaf Wasa* and preceded him as skipper on the *Thor*, and now monitors the consecutive masters of the *Atlantic*! (Figure 30).

---

[16] In the present case, the Maritime Law of 12 June 1891 was applicable.

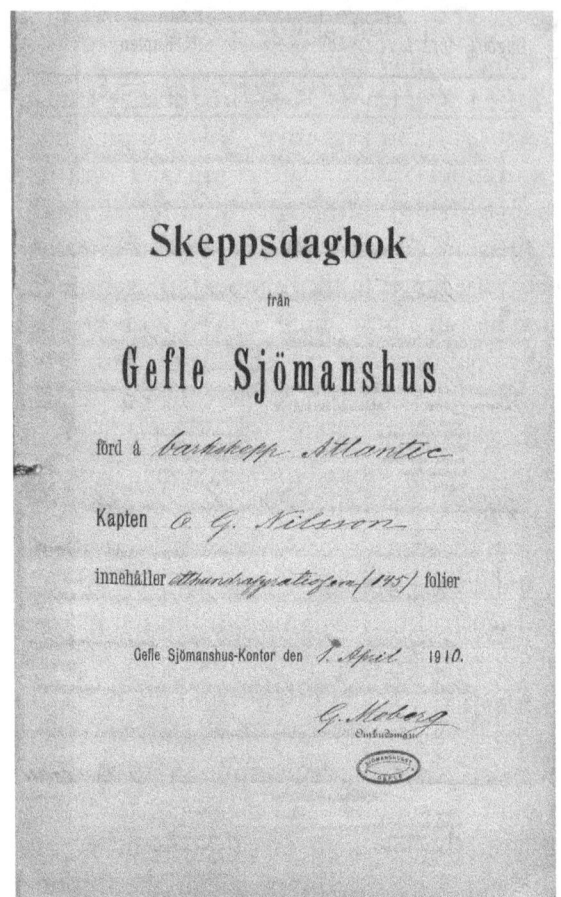

Fig. 30. *The last logbook of the Atlantic. Cover Page.* Private copy.

The spotted blue cover contains 145 lined folio sheets of which No. 132 is the last one used, responding to pages 266–267, where our story ends.

The master mariner was responsible for the journal. He could delegate the bookkeeping to the chief officer under supervision but had to be prepared to present it to consuls in the first place, in addition to harbour captains, maritime administrations, and shipping offices upon request, to have it checked and occasionally signed. An incomplete or missing journal was almost like a missing ship; at least it was the mark of a negligent or incompetent captain, who then was fined with 200 kr and could be dismissed. For long, the text was handwritten in ink with a steel pen dipped in an inkwell, and amendments and erasures were crimes. Normally, lost logbooks indicated very serious calamities on the

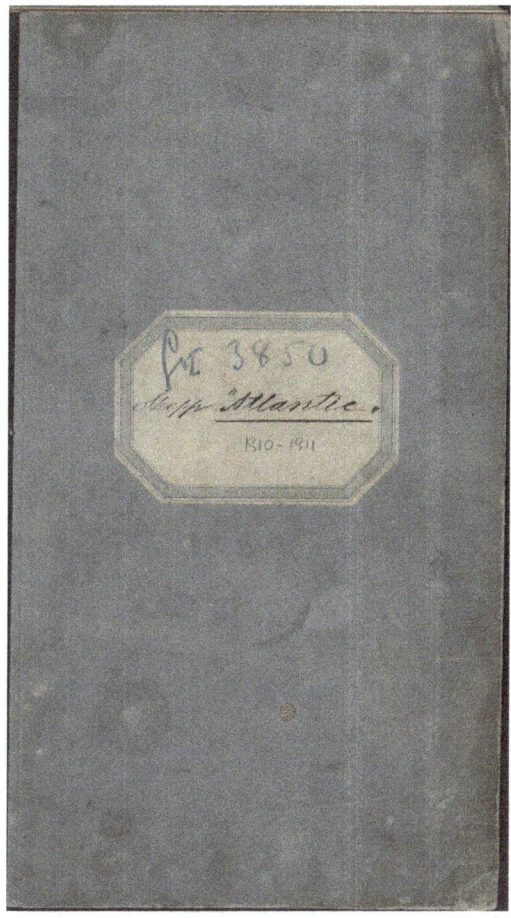

Fig. 31. *Cover of the last ship's journal.* Private copy.

sea. When a ship foundered, the master was the last to leave the vessel, carrying the journal in the flight. But he was not allowed to bring it home unless he was the owner of the ship (Figure 31).

11. Load Line and Waterline

The diarist on the *Atlantic* starts his entries even before April 1910, and the minutes give a unique glimpse of the strenuous and drawn-out work before all departures:

> Wednesday 23 March came on board with the mate Olof Thomasson [the author of many forthcoming journal entries],

along with constable H. Thomasson, carpenter O. Eneberg, Steward W. Thorsson, boatswain N. Tufvesson, and the able seaman E. Nilsson. Made the cargo spaces ready for loading incoming goods, and started rigging up. Maundy Thursday, 24 March, at 6 a.m., harbour pilot came on board, and we relocated to another, designated anchorage and moored there according to the harbour pilot's instructions. At 9 o'clock began to load and took in cargo until 6 p.m. The loading is executed by stevedores from land. Received barge No. 1 which contained 6,012 = 2 × 4" [laths], and barge No. 2 which contained 4,107 of different dimensions, both belonging to the First batch. The crew performed all kinds of work. Good Friday, 25 March, was kept free. Saturday, 26 March [Easter Eve] loading continued from 6 a.m. until 3.30 p.m. The crew did all kinds of work in the rigging. Sunday, 27 March [Easter Day] was kept free.

The three-week loading time has begun. Each day from now on, barges arrive from the shore with planks, some 90,000 pieces in all, in addition to a number of boxes, consisting of deal of various dimensions, packed in bundles and delivered to the ship by boat. The crew is busy stowing from 6 a.m. to approximately 6 p.m. six days a week, combined with making the ship ready for sail. 'March 31, the deckhand Wictor Svensson, deckhand E. Jonsson, and deckhand John Berggren have started working on board, also deckhand Skoglund'. Barge 7 and 8 arrives; 'loading from six to six. The crew sets sail and continues preparations. Friday, 1 April harbour pilot on board at 4 p.m., took home the moorings, and was towed out on the roadstead, anchored with port anchors and 30 fathoms of chain'.

Tuesday, 5 April loading from barge 16; the crew paints outboard. Wednesday, 6 April loading from barge 19; the crew painting. Thursday, 7 April loading 06–5.30 from barge 20 and 21, 'constituting the beginning of the end of the consignment' the captain notes with an almost audible sigh. The crew continues to paint externally. Friday, 8 April loading 06–5.30 from barge 22, and the crew performs various jobs as usual. Saturday, 9 April loading as usual 06–5.30 from barge 23. The crew scrubs inside. The wind varies. Snow fog in the morning, the barometer shows 29.9 [in mmHg].

Sunday, 10 April, fresh NE winds with clear air. At 8 a.m. measured the water level in the ship which was found to be 22 inches. Pumped dry. Monday, 11 April loading 06–5.30 from barge 24, 25 and 26; the crew continues to scrub inside. The water level in the room measured 20 inches. Tuesday, 12 April loading, the crew painted and scraped inside. Wednesday, 13 April, loaded all day and scraped inside. Thursday, 14 April loading from barge 31, hoists up the anchor. Friday, 15 April, the crew loaded all day, water level in cargo hold is 23 inches.

Saturday, 16 April: was loading to 05.30 p.m., received barge 27 with 13,389 pieces and then barge 28 with 160 boxes. Getting ready for deck cargo too! Barge 29 arrives with 2,105 bundles from barge 30, deck cargo is added. The crew painted, and the water level in the hold is 20 inches in the morning and 20 inches in the afternoon. Loaded from 6 a.m. to 5.30 p.m., the pilot came on board, towed the ship out to Gråberget to take cargo there, loading 2,018 bundles and from barge 32, 1,019 pieces. Water level in the hold: 23 inches, in the morning and the afternoon. The crew painted and scratched inside. Barge 33 with 570 bundles. Ready to sail!

Sunday, 17 April 'had a new inspection for which the whole day was used. Monday, 18 April: scraping the ship and getting ready to go. The ship lies on the load at 21½ feet'. This means the ship was deep draught to the limit. The surface of the outside water is almost on level with the inward deck. This is heavy and cumbersome shipment, tens of thousands of bundles, boxes and deck cargo, which may stabilize the craft to some extent but also make it creak and leak in the long joints of wood. The captain nevertheless observes:

> The ship is properly manned and in seaworthy condition. Tuesday, April 19, pilot came on board. Weighed anchor and went to sea. Left the pilot of the tugboat outside Eggegrund [lighthouse island off Gefle at N 60°43,740′ E 17°33,252′]. Hoists spanker sails.

At 5 o'clock, an inspector came on board to observe the water level and put control seal on the pumps. 09.00 Pumped dry. The wind varies from 10 a.m. The crew is painting inside. Calm and fog. Clearing outwards, the ship is well equipped; tug arrives to take us to sea. Set top sails, foresails and stay sails. Destination: London!

Tuesday, 19 April observations: course SE, barometer 29.4, variable winds. Wind direction, course, leeway, and patent log with nautical miles are henceforth carefully checked, calculated, and registered, as well as warranted soundings with the lead. However, there is no measurement of windspeed, which indicates that the old anemometer is gone, and the column 'knots' is left blank throughout. But the helmsman knows what he is doing. 'Beating to windward, standing on port and starboard tack respectively. Wednesday 20 April. Beating in variable winds, standing on port and starboard tack respectively'.

The passage to London goes without a hitch. The procedure is the same as for all previous years, yet it is interesting to see the elegant manoeuvres of the ship in sailing to windward, with the variability and special capacity of all different sails from stem to stern applied. For instance, even the oblique and lateral mobility of the square sails alone prove that merely very hard 180° headwinds were real opponents. Old navigators did not fear them either, but wisely avoided them by adjusting the course. The problem of a ship of advanced years is not navigation but stamina.

12. The Appearance of the Longshorewomen

From London they go to the Mo Steam Mill at Norrbyskär islands in the Gulf of Bothnia, an industrialized village with a harbour on the Norrland coast (Figure 32). The old shipping destination Mo does no longer exist, but its mill, timber yard and port for export was the beginning of the Mo-Domsjö timber industry (MoDo), now named Holmen, one of Sweden's largest forest owners and wood industries. One can have no doubts about the nature of cargo and loading time here, nor the ample use of lighthouses and pilots in the archipelago where it was difficult to navigate, it was the same as always, but we are provided with a drawing shufti of the exceptional residents working in the harbour.

With both anchors out on 30 fathoms of chain, and hawsers secured in the stern, unloading of ballast starts at once on Wednesday, 29 June 1910 and goes on to Saturday, July 2. Then twelve female stevedores immediately appear and start the loading of goods fore and aft. After the weekend, the whole workforce of dockers in the berth is mobilized in the loading: '*27 kvinnliga stufvare,*' ('twenty-seven woman

V. The Last Days of the *Atlantic*   249

Fig. 32.   *The Mo Harbour, a familiar destination to the ATLANTIC, now no longer found on nautical charts.* Copy in the Hallengren Archives.

stevedores'). The next day they are one more. This goes on in the same fashion the whole week until Saturday morning, 'when the women gave up at 8 a.m. due to the rainstorm'. But they return next week, and the next, and go on until 20 July, when they disembark and the *Atlantic* puts to sea with a draught of 20 feet fore and aft. Destination: London, again. You begin to feel the rhythm, don't you?

The unisexual crew returns to London with no female passengers and arrive at Gravesend on 11 August. All's well on board. They moor at a designated berth in the old East India export dock in Blackwell, and the lengthy process of unloading and reloading begins. On 7 September, they are all set, are towed the following day and finally leave the tug at Gravesend and set sail for Stockholm.

As they are steering northeast, the NE September winds they are facing are most unfavourable, and they become battered by gales, for some time have to bring to and ride at anchor. There is an alarming note of 16 September on launching starboard ship's boat indicating that danger is imminent and a lifeboat kept on standby. On 28 September, they reach Øresund (the Sound) and the Swedish west coast, and on 3 October call at Stockholm after a great deal of worry and trouble. Unloading was finished on 17 October.

The next day they proceed to Gefle, being towed almost the whole distance along the coast. In the homeport, the loading of woodware starts all over again—with female stevedores! Certainly a most welcome reunion, for the sake of variety. Thereafter, on 19 November, they are ready to leave—for London again! However, the North-Easter has been replaced by a South-East wind: adverse wind again! Winds are variable, however, and at times the party have light winds and hardly any steerageway at all. They make it, though, and reach London in December 1910. On Monday, 5 December, they cast anchor in Regent's Canal. Then follows the extended period of discharging, to pump dry and replenish the stores, and loading an unwieldy pitch-black and sooty cargo of coke.

They have their own way of celebrating Hogmanay! New Year's Eve, Saturday 31, is a working day in the dense fog, and so the New Year's Day, 1 January, when the fresh-water reservoir is refilled and the ship is towed to the Coke Company in Beckton, well-known for its Gas Works. Whence comes the solid fuel, whoopee! Nonetheless, on 4 January they are ready to go to Elsinore, Denmark, with a draught of slightly over seventeen feet. 'Headwind and fog'. The North-Easter is the main obstacle. At anchorage till 6 January. Safe and sound at the destination four days later, but is ice-bound! Partly new crew. Works in the rigging. Departure for Härnösand in the Baltic Sea 24 April, arrival on the 29th. On Saturday, 20 May, the *Atlantic* leaves for Hartlepool, England. Noteworthy is the ample use of lighthouses in taking bearings, and the need of towing in the Sound and inside the belt of skerries.

Arrival to Hartlepool on 9 June, waiting on the roadstead for pilot and high tide. No misadventures here. The unloading and loading procedure goes on all days through, including Midsummer Day, Saturday, 24 June 1911. When discharging is finished, a new coke cargo arrives. All crew is working nonstop. No women in sight in England. All work done on 7 July! Draught slightly more than 19 feet. Departure for Luleå in Norrland the next day. Strong breezes, fresh gale in the Baltic Sea, 16 July. Making land on 24 July.

Same procedures. Leaves for Storfors, Piteå, to take timber cargo to Hull, England.

Friday 11 August: starts loading timber with the help of female stevedores. On Saturday, full workforce: twenty woman dock workers. On Tuesday, twenty-four women, the next day 25 longshorewomen. The work is getting on fine. They are almost finished in the rainstorm of Friday, 18 August, when they weigh anchor to pick up more cargo at Hufvan, an island off Piteå in Norrbotten. It is the same kind of workers there to assist them: 'Monday 21 August loaded all day with the help of 24 women'. As in all previous ports, the ship's journal then notes *ditto* day after day. The monotony means that all goes well. On the first of September, 'lying at 21 feet aft and 20 feet six inches ahead', and completely seaworthy, as far as the officers can see, they decide to go from Piteå to Hull in this Swedish–British exchange, reminiscent of the barter economy of ancient times. The value of forests in the north measured against the value of collieries in the south! A win-win situation with transactions to mutual benefit.

13. Perseverance to the Utmost Limit

On 2 September, they depart southwards in the northerly wind. Navigating by means of the number of lighthouses on the way, they proceed fairly well the following days, while the wind varies fickly both in direction and force. This makes them utilize effectively the full proficiency of ship and crew to alternate between beating to windward, tacking, and run before the wind. Passing eastward of Fårö and Gotland in the southern Baltic Sea, almost mid-sea between Sweden and Latvia. *Real* bad weather sets in 9–10 September; rising to storm. The sails are secured except the lower topsail in the mainmast, which is broken; a veering manoeuvre has taken its tolls. The interior water level rises to 29 inches, the wind pump is constantly on in the hard, yet gusty winds. The captain notes: 'the ship is working hard'. The water level rises to 36 inches.

After Skanør–Falsterbo (southwestern Scania) in the Sound the situation becomes awkward and alarming when the storm has abated, as the wind pump is not working anymore in the calm. How much water the vessel makes, becomes clearly visible. The crew is occupied by keeping the handpumps going, working ceaselessly at all hours. 'The ship leaks four inches every half-hour, so it is decided to anchor in the Malmö roadstead to examine the ship'. The water level is 30 inches downstairs. There is hard work pumping dry, occasionally in assistance by the windmill—when the wind is blowing. Friday, 14 September 1911, the captain goes ashore to notify the shipowner and the cargo owner about the situation. On Friday, 15 September, the water rushes in.

The cargo is brought ashore and stored on the quay; anchors and chains are laid out from the ship to lessen the draught and the flow. Two hands escape the ship: Laurits Jensen and Olof Olsson. The ship is inspected by the port authorities. When the burden is lightened, and the leakage in the stem is over the water line, it can be appraised and repaired by a harbour carpenter assisted by the ship's carpenter.

On 20 September, the reloading starts, and next day both repair work and recharging are finished. The case of the *Atlantic* has been duly taken to court for trial at Malmö Rådhusrätt, the municipal court. The judicial decision is favourable, and the ship and its master obtain a certificate of seaworthiness on 22 September 1911. The water level in the vessel is then 16 inches; they pump her dry. The crew replenish the stores of provisions and water. The ship now leaks only '10 inches in 12 hours'. What a relief! On 24 September, they weigh anchors and take in the 25 fathoms of chain in the roadstead, and continue the trip to Hull, moving from Latitude 55° to 53° after the long roundabout north through the Sound and then south through The North Sea.

On 3 October, they are safely arrived after having mastered very changeable winds, and are comfortably off, being towed on 5 October to Albert Dock. This is Kingston upon Hull in East Yorkshire, a historic and beautiful city situated at the mouth of the Humber River, 80 km east of Leeds and 55 km southeast of York. There is no time for sightseeing. Unloading 6 October onwards. On 2 November, relocation to another berth to take in 60 tons of ballast on top of 250 tons of sand. On 3 November, lying at a depth of slightly over 14 feet in stem and stern, the captain announces that the ship is seaworthy and ready to sail upon the sand, and they leave next day. Destination north: Sundsvall, Norrland, for more timber cargo!

14. Fighting to the Bitter End

Heading northwards, they are met by increasingly harder winds, rising to storm, and rain squalls. The crew is busy in the rigging, continuously adjusting sails according to the challenges. The master mariner also notes down on 5 November that 'the ship is working hard'.

6–7 October: hurricane. 'Westerly wind increases to storm and hurricane strength, impossible to scud and run before the wind; we bring to and lie at half-reefed lower top sail in the main mast [...] Both hand pumps and wind turbine are running at maximum speed...'

Large waves hit the hull hard; high seas wash the ship, and a huge wave from larboard strikes and causes vast interior damage.

Now the aim is to get safely out of the storm, pump the ship dry around the clock, and steering for the Skagen Lighthouse, on the northernmost peninsula of Denmark—known for the contemporary Skagen painters and their light-hearted life on this sunny and beautiful place.

On 8 October 1911, the *Atlantic* has reached the Sound, and finally manage to put into Prince Hamlet's Elsinore, Denmark, which becomes their port of refuge, and their destiny.

On 9 November, deckhand Artur Madsen, who was injured during this trip, is signed off.

The captain goes ashore to inform the shipowner of the extent of the damage. A strong gale prevails.

10 November, ditto for anchor.

11 November, inspection of the ship.

16 November. Condemnation of the ship:

'On the occasion of the Condemnation of the Ship, this Logbook is hereby cancelled'.

I anledning av Condemnation af Skibet annulleres herved denne Log bog.

Kungl. Svenska Vicekonsulatet i Helsingør d. 16/11 11. (The Royal Swedish Vice Consulate in Elsinore on 16/11 11.)

The mixed language shows that the Swedish consul is Danish born. The book is closed and sealed, and the itinerary terminated. It is the end of time, and all is over.

15. The Wreck

In its existing condition, the *Atlantic* was purchased by the shippers Öberg & Horndahl in Helsingborg across the Sound. The wreck was finally sold to Captain Johan Åkesson in the same seaport. It was then brought to Torekov, the old fishing village and famous summer resort, which according to the local legend was named after a girl, later known as Saint Thora, who was drowned by her stepmother.

For the *Atlantic*, the final break-up of 1911 was at hand. The day of destruction.

Sic transit gloria mundi, 'thus passes the glory of the world'.

However, you may rest assured that most parts of the body were recycled and went into other constructions along the coast! Not a splinter was preserved as a mere matter of memory. Wrecked goods were valuable to the handy, the oak in an old hulk a treasure for builders and carpenters, and firewood welcome in the unwooded western part of the Scania province.

In those days before consumerism, people took charge of everything, and there was little refuse. Throw-away articles were not invented yet. Sweaters, blankets, watches, bedsteads, household utensils, fishing-rods and nets, rowboats, and furniture—in the countryside solidly built by hardwood to last for hundreds of years—were used by generation after generation.

When broken up, wooden ships were transmogrified into shackles, stakes, piles, huts, flagpoles, floors, beams, boathouses, dinghies, fixtures, furnishings, seaside fishing cottages which are today high in price; crofts, foundations, grit bins, garden beds and railings, ground filling, decorations, art. Steering wheels embellished the ceilings and became fittings.

Nautical instruments were carefully preserved, mended and reclaimed, as were lights, lamps and other brass objects. Sails and cordage of hemp and flax as well as tarpaulin got new ranges of application. In certain respects, there was an increase in appreciation. In reality, these ships never vanished. Use, material, and value lasted.

VI. What Happened Next?

1. Wartime

NEXT YEAR, EMMY'S OLDEST SON SVEN—THEN ALREADY a 17-year-old upper secondary school student—was permitted to accompany his father Gotthard on a trip to Stockholm to attend the 1912 Summer Olympics. This was an iconic multi-sport event in the newly built Stockholm Stadium, particularly memorable to the host nation because Sweden won the most medals of all and emerged a winner. Unforgettable was also the US general Patton's participation in the pentathlon; the competition in literature (!) won by the organizer Pierre de Coubertin; Shiso Kanaguri, the Japanese marathon runner, who mysteriously disappeared in the middle of the race, eventually to turn up and complete the distance fifty-four years later…

The wrestling competitions were also outdoors and went on to back fall. The longest match was the semi-final in middleweight A between a Russian and a Finn. It lasted for 11 hours before it was broken at nightfall. None of the exhausted combatants made it to the final match the next day, which was won on a walk-over by the Swede Claes Johansson, who was very happy—as was the jubilant audience!

One of them was father Gotthard, who stood up cheering and shouting in all the finals and never wanted to stop. His teenage son was a little ashamed of this. The two are glimpsed in a newsreel where the son is seen in the stands in a strikingly elegant jacket. Two years later, he was a graduate and at once recruited and deployed during the First World War, 1914–1918, and was in military clothes practically until he, as a non-commissioned officer, attracted the Spanish flu, the 1918 influenza pandemic—a virus that killed more people than the war—an illness from which he narrowly escaped alive. His old army hickory skis are still preserved. Also, the letter of official exemption from military service that finally came from King Gustaf V in the Second World War, when he was a widower and single parent of young Jewish daughters in the time of Nazi German siege and occupation.

These horrible years of war, famine, decline in trade and shipping, mined waters and shortage of everything, struck Emmy's sister Maria and her family miserably. Moreover, Mia never overcame the tragedies of her life, and these dark times made things even worse. She mourned the death of her beloved John (in 1901), the sea captain she had met in Australia, whom she had married of ardent love in faith, hope and charity; joined as seafarer worldwide and who now rested eternally in a South African cemetery on the other side of the globe. The oceans were her abode and true home, and the origin of their offspring too—Axel, Barbro, and Gurli—who had all been to sea.

She hated to be alone and lived for her children, and when her father, Captain Söderström, died in 1902, and her daughter Barbro passed away in 1903, she was increasingly protective and desperate. She harboured a desire for love and family, and a longing for the sea and the glorious Australia, a longing that her son would inherit.

In this mood, she married a travelling salesman in 1905 who caused her many sorrows. She led a roving life with him, as he did on his own, and before he died in 1919, he had ran through most of the money she had inherited from her husband and her father. It was an outrageous and chaotic marriage.

Her quick-tempered son Axel, loyal to his mother and in disagreement with his stepfather from the start, after a brawl was turned out of doors into the snow on the first Christmas Eve together with him. After that the couple saw no more trace of the son. Idolizing his father, he wanted to become a seaman. At 15, in 1904, he had enlisted in the Härnösand mercantile marine office, having reached the age limit for enrolment. He became a wanderer at 16 and sought odd jobs in the harbours and on boats. Later he was hired by runners as a casual labourer and deckhand, and he appears to have travelled with various foreign vessels from port to port and jumped ships in different harbours abroad, free as the air. He was a migrant.

According to his own version, intimated to his mother when he returned in 1919, he ended up in Australia at the outbreak of the First World War, by then a point of no return, and his escape in the foreign port was not reported. This kind of seamen were more common than any statistics can show, obviously, but this also means that his life story and whereabouts cannot be verified with certainty. Numerous foreign sailors jumped ships in Australia, few of which were officially reported to shipowners and to the police, and many of them changed their names. Furthermore, there were numerous wartime refugees. These facts below, on the other hand, are well-known and evident.

John Stanley Martin, of the University of Melbourne, once compiled an impressive 'Register of Ship Deserters, Melbourne 1878–1924'. This was further developed and analyzed, with a focus on Swedish seamen, by the founder and director (1966–2002) of the Swedish Institute of Emigration, Professor Ulf Beijbom. From his research, it transpires that in 1885 alone 215 Swedes 'jumped ship' in Melbourne, most of them from British boats [Beijbom, 1983, 229]. Then again, these are merely the reported and registered cases, and only of one harbour.

Australia was known among Swedish sailors, workers, and outcasts as a place where you could start anew with a fresh identity, and where nobody was asking for *antecedentia*. Like America, allegedly a similar country of opportunity, it was giving a second chance to everyone! To adduce a single example to illustrate the problem facing us in our investigations, the naval historian Ingvar Henricson spent four years tracing Jan Erik Andersson (1860–1931), a sailor from Hållnäs, Sweden, who escaped in Melbourne in 1884 and never returned. The key was that his name was not Jan Erik Andersson in Australia but Albert Bernhard Anderson.

Concerning our man, Axel Erik Douglas Ljungberg, born in Westervik on 24 July 1889, it can be stated that he does not appear in any Australian registers either. There are indications that he sometimes spelled his name differently, called himself Edward instead of Erik, or John (his father's name) and occasionally Youngberg—as another Ljungberg relation did: Emy's daughter-in law's nephew, Raymond Ljungberg, who as Ray Youngberg was a sheriff in America. Swedes were an emigrating people, more or less adapting their names to the new environments in America and Australia.

Apparently, to get even the slightest idea of what actually happened to Maria's son, we must start elsewhere, and at any rate we will then learn something about the setting and the mindset of the period.

The aim of the Entente powers—Russia, France and Britain—was to cause a breach in the six hundred-year-old Islamic Ottoman Empire by taking control of the Turkish straits, the Dardanelles and the Bosporus, which connect the Aegean Sea, the Mediterranean and the Black Sea via the Sea of Marmara. Hence, the object of the 1915 campaign was to secure and open up all waterways connecting the Black Sea in the east with the Atlantic in the west—and the Mediterranean Sea with the Indian Ocean through the Suez Canal. A strategy as plain as pikestaff on the drawing-board. Consequently, it was all about navigation and naval control.

Similar strategies were drawn up by Carl von Clausewitz in the Napoleonic era and particularly by Admiral Sergey Gorshkov in the Second World War and further on in the Cold War, and more successfully implemented by them than by First Lord Winston Churchill and his commanders of 1915. The battle of Gelibolu (Gallipoli)—from the Greek Καλλίπολις, meaning 'beautiful city', incidentally the name of Plato's utopia—ended up one of the most catastrophic defeats in the First World War, and the circumstances were entirely chaotic. The outcome was half a million dead and wounded in total, both sides suffering horribly. Yet, it became the victory of Kemal Atatürk, the founder of modern Turkey, which at this time began its pogroms against Christian Armenians and other ethnic groups, adding to the innumerable war causalities worldwide—to no avail but horror.

In the British world Down Under, the campaign in the Dardanelles, the Strait of Çanakkale, had meant the forming of the united Australian and New Zealand Army Corps (ANZAC), and almost panicky conscription of volunteers and seamen, both nations being dependent subdivisions of the Commonwealth. In Australia, the sixteenth Battalion was raised in September 1914 as part of the *all-volunteer* Australian Imperial Force for service during the First World War and assigned to the fourth Brigade. After completing short training in Australia, the battalion first embarked for Egypt. Enrolment of volunteers for various services continued in 1915. People of many nations, men and women, volunteered for large numbers in auxiliary roles. The battalion joined the New Zealand division for participating in the fatal Gallipoli campaign.

Many of these recruits were badly trained for military manoeuvres and combat. So was the sailor and labourer Ljungberg and thousands of other volunteering expats. He said that he and other friends enlisted when they had a drop too much, to impress some girls, and that this all came to a disastrous end. Horrific!

Instead of bragging afterwards, he was ashamed of himself and thought he was a failure. He admitted that civilians of combat age were harassed by the women who had their husbands and brothers forcibly conscripted, and that stranded sailors felt compelled to enrol. He was no soldier, no hero, never participated in pitched battle, but was terribly injured in an accident on the sea during a transport, perhaps in the minefields.

In the early 2009, the present author, then working in New South Wales, had the opportunity to investigate Axel Ljungberg's whereabouts and fate in Australian sources, and had appointments with heads of the RSL (Returned and Services League of Australia) and the pastor of the Church of Sweden in Sydney, who searched all files that summer. The combined effort of both institutions meant found the man. I was honourably invited to the headquarters of the League's New South Wales branch at 351 Bay Street, Brighton-Le-Sands, as it was considered that a cousin of my father was an Australian ex-service man and belonged to the recruits that are celebrated at ANZAC Day in April each year and are monumentally memorized. As a researcher, I entertained doubts, however.

The name was spelled a bit differently, though, and his age was not stated quite accurately, but these inaccuracies did not surprise or bother any of the experts I talked to. It was often like that, and given names often differed. To enrol was like taking on a ship in the last minute, and people were much in demand. All personal information about origin and experience was otherwise as expected. Copies of all papers were produced.

They stated that Mr Ljungberg joined the sixteenth Battalion of the Australian Imperial Force on 25 January, 1915. Furthermore, that he was an unmarried Swedish man with naval training, working as a wharf labourer, with his next of kin a woman in Port Adelaide, and was prepared for inoculation against smallpox in the service. So far, so good, but the army papers ended with a sheet on 23 March 1915, where he is relegated from the fifth reinforcements of the sixteenth Infantry 'for bad conduct'. Another concluding item is marked 'discharged' in red.

According to the family story, passed on by his mother's well-informed grandchildren [Herrström 1973, 2009], Axel Ljungberg was, as seaman and labourer, in the spring of 1915 recruited as mariner in the volunteer corps handling transports, serving as worker and deckhand, not in military service. He was transferred to shipping, and one way or another ended up as supernumerary in the theatre of operations. He was terribly injured, probably in one of the number of wreckages and accidents, and the family well knows that he afterwards was treated at a hospital. The ships with which he returned home, in an emaciated state, are unknown, as are the particulars of his foreign experiences. He did not brag of any deeds of valour, or pleaded to anyone for pity, was down and quiet, only intimating what had passed to his dear mother, the circumnavigator, now living in Stockholm and widowed for the second time.

To her he fled as an inconsolable child in the port of departure and arrival, maintaining a wall of silence to visitors and friends. However, the fact that his body was covered by horrible scars he could not always hide, and these injuries upset his dear ones, and secretly testified to a fate that was evidently not all made up. His forebears of three generations as well as his parents being seafarers, and his surviving sister named after a ship, he kept up the tradition of adventure and hard work in his own way. He shared the family misfortune—after the tragic demise of grandfather, father, stepfather and sister—and was now disabled, unemployed, infamous, evasive, an addict, and was dodging all questions. He was known as a strong and rowdy combatant, once throwing an opponent or two out through a window. That may confirm the note on bad behaviour.

As a broken man he thus returned to Sweden as a mere shadow of his former self, dispirited and taciturn. After a turbulent life as a sailor, and finally, during his last years, leading an increasingly lonely and anonymous existence, he died at Norrköping on 29 October 1929, just 40 years old. He never married, but rumours in the family indicated that he might have left a fatherless child in Australia with any of his old female acquaintances alluded to, and which is also partway implied in the protocols. His mother thought she had a grandchild in Australia.

Late in his life he did return to the sea, however, and the family legend of an American naval license may corroborate (or stem from) the fact that he in 1926 was employed as able-bodied seaman on the Norwegian steamer *Modig* of Ivar Christensen, Oslo, commanded by Captain Lars Handeland, and destined for the USA. This is the last note about him in the shipping records. As a seaman he died with all flags flying, in his own right a transmitter of an ancient tradition. His mother arranged for his interment in the Westervik memorial, adding his name to that of her father, Captain Johan Axel Söderström, and her daughter Barbro.

2. Ghosts

Mia's only surviving child, the daughter who was named after the sailing ship *Gurli*, and as a baby had been attacked by an eagle or condor onboard, henceforth were Mia's only subject of rejoicing and hope and served as a ray of sunshine. A steady, multitalented, handsome and lively person with a rippling laughter who at her brother Axel's death had already given birth to three children, including the two mentioned genealogists and mathematicians, Yngve and Stig Herrström, and whose grandchildren became artists, writers, entrepreneurs, and all very gifted. Gurli carried on the shipping trade of the family as secretary of the shipping company Svea (Rederi AB Svea).

Grandma Mia had been faced by another kind of heritage than talent. It was one of sombre and tragic nature, and this she shared with a whole generation. In a way, this reminded of the irresistible evil Fate, and the vicissitudes of Fortune, staged in the classical drama, and for thousands of years had been connected with Nemesis or divine retribution as the consequence of hubris or transgression. Yet it was as elusive, capricious, involuntary and impenetrable as the attack of an insidious plague of unknown origin.

Her sister Emmy had also seen the workings of this dark force, and everyone knew about this invisible threat, familiar to people busy in trade and shipping and international travel. Sin, shame and guilt were involved, and yet there was hardly any protection or cure against the enemy. In death certificates, the doctors noted *lues*—the common curse in the age of sail since Columbian times.

In Mia's and Emy's days, the female sex was the scapegoat. The nineteenth century seethes with the mythic image of the *femme fatale* and *la belle dame sans merci* who seal the doom of incautious men, demoralized by beauty and seduction—ideas derived from the Medieval concept of the pudenda as the gateway to hell and notions from the age of witch trials. However, it takes two to tango, and in reality women were primary receivers of syphilis and furthermore harboured the bane of congenital syphilis, which was passed on to their foetus. In the 1880s, the world-famous Norwegian playwright Henrik Ibsen pressed this sore point in 'Ghosts' (*Gengangere*), first staged in Chicago 1882, showing the infernal venereal disease as an inherited sin which in disguise is passed on through generations, always detected late. This social disease was a family affair.

The illness could pass unnoticed and be latent, but a decade or so after the initial infection it might reappear in its last, awful, tertiary form with disastrous consequences—as in *paralysie générale*, which was the term used about this devastating dementia in their time, mental derangement. The Söderström, Ljungberg, Ström, Videnstam and Hallengren families all had witnessed this horror and felt the fear of the diagnosis and the complications.

There was no remedy available except—perchance—continence and the extinction of the human species. But many deep-sea sailors had observed that they were cured from their ailment after having taken ill with the endemic tropical fever (malaria). This meant replacing one evil with another, but the measure answered to a conviction among nature healers and homeopaths of the antagonism between afflictions. During the First World War, the Austrian neuro-psychiatrist Julius Wagner-Jauregg with significant success tested inoculation of malaria parasites in sufferers of neuro-syphilis, for which he was awarded the Nobel Prize in Physiology or Medicine in 1927. A comparable relief to the long-suffering populations who were poisoned by the predominant miracle cures of mercury and arsenic!

In fact, old people remembered this shivering fit of ague also! Malaria fevers had been common in Swedish coastal cities, particularly in the eighteenth century. Carolus Linnaeus, the famous botanist, was familiar with malaria from his hometown—there known as the Upsala fever—and presented a doctor's thesis about its cause (*De febrium intermittentium causa*, 1735) suggesting quinine as remedy. Sea transport networks expanding in reach and volume of passengers and goods carried, implied that pathogens and their vectors could move further, faster and in increasingly greater numbers [Tatem et al, 2006]. When it was reported in the harbour that a sailor—to great amazement!—had died from natural causes this was often understood as an allusion to venereal or other infectious diseases.

The number of plagues spread by boats and their crews, passengers and other world travellers is startling and constitutes the ultimate drawback of international trade and shipping. This has set deep marks on our civilization in the five hundred years of post-Columbian exploration and voyaging, and prostitution was booming in pace with the Age of Sail trading of colonial products and the surge on export and import markets. Contagions flourished in the contact period of the new era when obscure fiends disembarked ominous ghost ships of distant ports—bubonic plague; cholera; the *Aedes aegypti* vector of yellow and dengue fever viruses, which the transatlantic slave ships brought from West Africa to Spain and Portugal via the Americas. As with the social diseases, the exchange was always heavy.

The fight against these invincible adversaries remained mere shadowboxing as long as they were invisible and unknown microorganisms, like for instance the spirochete *Treponema pallidum*, the spirally twisted bacterium that causes syphilis. Into the bargain, the infection could pass under the radar for years as we have seen. On top of it then, this treacherous foe had a final sinister sneak raid in store: the devastating quaternary neurosyphilis—dreadful as a gas attack; a weapon of mass destruction in its time.

To cap it all, this abomination came as a bombshell in poor Maria's ill-fated second marriage with delayed release. This was a tragedy beyond the reach of drugs, appeals and prayers and there was no redemption in sight. Her husband's illness struck mercilessly. He got out of senses, lost his reason and health irrevocably. He suffered badly and wrong, and passed away in dementia at 54, fourteen years after their wedding. A gifted, temperamental and venturesome *bon vivant*, a scarcely successful businessman and sales representative with a zest for life, who was paired with a clean-living and philosophical woman who had never been outgoing, never danced, read American philosophers like Theodore Parker and William Ellery Channing, pondered over the human condition, delighted in poems and hymns, sought illumination by freedom from sins, and to whom debauchery was obnoxious. Yet, undeniably they were lovers, and her passions were strong. An ill-matched couple in a matrimony where it took two to make a quarrel. Maria's last, longest and childless love affair was so sad. Ibsen's play does not suffice to illustrate the setting. It requires August Strindberg's *The Dance of Death* (1900) or Edward Albee's *Who's Afraid of Virginia Woolf?* (1962) to stage it.

And no wonder then that Mia henceforth hung on to her only source of hope and joy, her dearly-beloved daughter Gurli, the only surviving child and family member. She would watch her mother with loving care for the rest of her life, which was not always an easy thing to do as the mother was anxious and difficult to please, all the more when the state of her health changed for the worse.

Gurli's fourth child, Ingrid Gurli Maria Stedman, née Herrström, born in 1932, in 2021 tells the present author on the telephone that her grandmother had a stroke the year she was born, and after that was partly paralyzed, feeble and impaired, and that her mother paid attention to her until she died. Ingrid furthermore divulges that the woman who wrote the circumnavigation diary was left-handed, was paralyzed on that side, that she walked with a cane, and had difficulty writing. That she stayed in different private homes, which Gurli arranged for, and that Grandma Mia for some time also came home and lived with Gurli and her family and played the piano; wanted the young Ingrid to play for her; that she was paralyzed until she died in Stockholm, fifteen years after she had fallen ill.

From parish registers and censuses of the population it comes to light that the itinerant Maria had moved between many different cities in her life, and in each place, there had often been several changes of address. After the years in Gefle (1867–1889) and Westervik (1889–1903) she lived in Norrköping (1903–1907), Helsingborg (1907–1908), Malmö (1908–1910), Höör (1910–1911) and finally Stockholm (1911–1947). The last part of her life is summarized succinctly in a note received in 2021 from the Archivist Åsa Lundborg of the Stockholm City Archives:

Maria Mathilda Charlotta Ström, born Söderström on 16 June 1867 in Gävle, died in Brännkyrka parish, Stockholm County, March 22, 1947. She moved to Hedvig Eleonora parish in Stockholm on April 8, 1911, from Höör's parish, Malmöhus county, together with her husband, the merchant Ture Edvard Ström, born May 24, 1865 in Alingsås, Älvsborg County, and her daughter from a previous marriage, Gurli Maria Ljungberg, born on September 23, 1893 in the Västervik parish. The couple had been married since June 11, 1905. Ture Edvard died on 16 November 1919 in Maria Magdalena parish, Stockholm. The daughter Gurli Maria Herrström, born Ljungberg, died according to Sweden's death register on 9 August 1960 in the Holy Trinity Parish, Kristianstad County.

The Maria Magdalena Parish Book of Deaths and Funerals in 1919 carries the appalling information about the demise of Maria Ström's husband: *Dementia paralytica*, Långbro mental hospital.

Her daughter Gurli had a strong family feeling and talked much about her relatives and the Gefle years, and had felt flattered and proud in her youth when an old sea captain in Landskrona, upon hearing her maiden name, exclaimed: 'Daughter of Captain Ljungberg of the *Gurli* and grandchild of the Skipper King, well, I say!'

Her family equally remembered how in her last days Maria sat surrounded by family portraits and souvenirs dreaming about the sea and her travels in the past; was rereading her own journal, recalling the far horizons and the jewelled sea of Australia where she met her love (Figure 33). Furthermore, how the old woman as a pietist dissenter, an evangelical reader in the modern freedom of religion, in her loneliness sang revivalist songs with raucous voice, and as always turned to Scripture in storm and doldrums, death and perils for solace in her despair and never lost her trust: 'Generations come and generations go, but the earth remains forever'.

Maria kept in touch with her sister Emmy, but they did not see each other often and did not write much due to the distance and the disability. Rooted forever in their birthplace, Emmy now lived in a different world, as it were, with a large family and establishment to manage, but her life was far from problem-free. A similar misfortune had happened to her firstborn daughter Greta, who had come to grief when she tried to be put in the family way. In 1917, she had wedded the shipbroker

Fig. 33. *Mia, the writer and traveller, in her old age among heirlooms, pictures, and memories.* Courtesy of the Maria L. Hallengren Collection.

Karl Vilhelm Adolf Videnstam. The marriage was childless, so the couple were beside themselves of joy when Greta was finally expecting in 1930.

As ill luck would have it, the pregnancy ended in a miscarriage. Next year, however, on Wednesday 23 December 1931, the day before Christmas Eve, she gave birth to the much longed-for child, a daughter. The baby was christened Kristina Margareta Videnstam, but with a narrow margin she survived New Year and passed away after twelve days life on Earth on Monday, 4 January 1932.

To her dear brother and confidant, Emmy's oldest son Sven, Greta disclosed that the stillbirth had a sore—the significant single chancre, the sign of primary stage syphilis. The foetus was infected. Syphilis had been transmitted from mother to child in both cases. Kristina was interred at the family burial place in Gefle.

A decade later, Greta's husband Adolf became a mental patient for his own share of the same illness that had reached its final stage—general paresis with dementia, personality changes, delusions, megalomania and paranoia. Greta's niece Maritha Karling called on him at the Ulleråker mental hospital where so many suffering genii had been dwelling, including the deranged artist Ernst Josephson. She found Adolf sitting anxiety-ridden on his bed with a handkerchief over his face, trembling with fear.

In the ceaseless question of responsibility, speculations blamed either an American woman's lips or an unwashed beer-glass; what ideas! To trace this contagion was impossible as this death-mill ground slowly. His wife might as well have been an asymptomatic carrier of the disease for a long time, likewise her Aunt Maria, possibly. It was a case of congenital syphilis. This revenant haunted the living for generations. Nobody could be blamed and there was yet no cure for this affliction although universal administration of penicillin—a discovery awarded the Nobel Prize of 1945—was in the pipeline. People were groping in the dark.

3. Family Affairs

Women of seafaring families needed a good share of valour and patience. They were given a baby when the sailors after several years came home for a while with malaria and syphilis, and the wives became widows early and the children fatherless and not provided for. Tuberculosis and cholera ravaged seaports at home. The families we have followed in this book carried on the tradition of trade and shipping by climbing the social ladder to the office level, facing other challenges as well.

For the family line we are now following to illustrate this thesis, there was a solid base at Kyrkogatan 17 in Gefle, the edifice erected after the fire of 1869 and housing the PJ Haegerstrand shipping office. The headquarters of this shipping company was familiar ground to Captain Johan Axel Söderström, whose *Atlantic* had sailed on its orders, and as deckhand he had sailed for the legendary Captain Frans Arnljot Carlson on its ship *Daniel* in 1860. Axel Söderström's family knew all captains and staff of the agency. As we have seen, his daughter Emmy married one of its clerks, Gotthard Hallengren, and for a while they actually stayed at the address.

Through his sister Aquilina, an artist, Gotthard was distantly related to the composer Jacob Adolf Hägg (her brother-in-law) and his cousin the marine painter Jacob Hägg, and was personally acquainted with them. As harbour glazier, Gotthard's father Jacob delivered the glass needed at the wharfs, so the childhood years had been spent among carpenters, shipbuilders and sailors, and as bachelor and apprentice he lived in shipbroker PJ Haegerstrand's house. Erik Axel Waxin, owner of the shipbroking company Haegerstrand 1909–1919, was the son of his sister Anna (1849–1887).

In 1919, Erik Waxin was replaced on the board of directors and owners by Greta's husband Adolf. Do you follow? This is only outlined in some detail to show that the family connection with this firm of shipbrokers, which started on the sea in the 1860s, has moved ashore to the boardrooms, *quod erat demonstrandum*. This illustrates both the professional continuation and the social mobility analyzed in this book. On a larger scale, this represents changes taking place in the middle class, movements of salaried employees, and also a new role for women in these social strata [Söderberg, 1972]. Looking more like the spirit of the times materialized than just a coincidence, this is the year of universal woman suffrage.

Formally, Adolf Videnstam was the owner of the shipping company from 1919 to 1945, but due to his declining health his wife Greta gradually took over the business, in practise she was the manager from 1941 (Figure 34). She was the owner of the company 1945–1957, in the latter part together with her brother Bo Hallengren, who succeed her as sole owner in 1957–1963. He sold the company in 1963 [Henricson, 2004].

Seen in this perspective, it was a family business for a century, although the seamen had become landlubbers. The generation of circumnavigators was succeeded by a generation of shipbrokers. Emmy's son Bo, head of the international Haegerstrand Shipping, competed with the contemporary shipping empire of Aristoteles Onassis, who died the same year, in 1975.

A few years earlier, the humorous shipbroker Bo at Haegerstrand Shipping & Logistics in Gefle brought the present author to see the busy Telex machine in his former office, and then recalled a message late at night from a sea captain who was delayed and his ship still remaining off the coast abreast of the old fishing hamlet Bönan: 'Staying the night at Bönan'. As the place name was also slang for 'the girl' he could not resist despatching a company approval of the nocturnal affair. There was no answer.

Fig. 34. *Emy's daughter Greta took charge of the leading shipping office and became shipbroker.* Photograph owned by the author.

He was full of fun and a storyteller, known for his sick humour, which gradually blackened. One of his last one-liners read: 'A man who had stolen a bicycle left at a cemetery railing defended his action by stating that he thought the owner was dead'. Another: 'A priest visiting a patient in intensive care received a slip of paper from the ailing man a moment before he expired, and afterwards read the message: "Don't sit on the oxygen tubing!"' Bo's own visits to the burial place became increasingly frequent as he wanted to 'get accustomed to it'.

His facetiousness turned macabre in later years and ever more morbid synchronous with the progress of his emphysema, which slowly suffocated and finally silenced him, but to the last days he played the violin with brilliance, his farewell music.

Greta, who carried the horrible venereal disease but enjoyed excellent health and upheld a positive attitude to life, survived her husband for ten years with a new man and died of Colorectal cancer at Upsala in 1967, and was buried in the same grave as her only child, the little Kristina.

Like her brother and all the siblings, she had a musical ear, and all of them spoke about the musical atmosphere of their parental home, which had set the tone of their lives and made them keep up their courage in all vicissitudes of fortune. Their sister Gunborg Emmy Maria Bergendal, who lived until 1988, had a good memory for details and communicated the following particulars.

After the turn of the century, Emy's children were to have a secret signal between themselves. If they got lost in the throng of people, or wanted to attract a sibling's attention, or if they were to congregate at a meeting place, they whistled a tune, the celebrated *Minuet* by the eighteenth-century Italian composer Luigi Boccherini, a famous melody from the third movement of his String Quintet in E major, Op. 11, No. 5.

The signal could be heard in the most unexpected places, after school, on Sunday, or in the street at night. In their home, chamber music was played, and the classical repertoire familiar. It was a large family, where all were musicians. The daughters Gunborg and Ingrid gave piano lessons for children in their neighbour Gustav Rainer's and Ella Schulze's home, one of them who was to become a cabinet minister.

Two sons excelled in violin, Bo and Sven. After the First World War mobilization and having begun university studies in Gothenburg, the latter was employed by the Grand Theatre in the city as violinist, but due to the general strike proclaimed by the Union of Swedish Musicians in 1922, he took on the steamer S/S *Stockholm* and went to New York as musician, and the refugee was therefore condemned indefinitely as strike-breaker by his colleagues and the official trade journal *Musikern* back in Sweden.

His mother, who had been pregnant twelve times and given birth to nine children, seven of which grew to manhood and womanhood, found various ways of supporting the household. She carried on a private employment agency for domestic servants; served as agent for a piano manufacturer; and played the piano at cinemas, giving voice and mood to silent film. Sometimes the film companies provided scores and instructions, at other times it was improvization throughout. When people were running on the screen, she made them run even faster; funeral marshes thundered as cliff-hangers, and in tearful romance she played with the heart attached to the strings. Times gone by!

The introduction of sound film in 1929, was a technical advance that left her unemployed. Yet the traffic at home in the Central Palace had continued, some visitors looking for a maid, others wanting to try out a new piano, so there were tinkling and talk all day. Furthermore, she was married to a harbour clerk and cashier who in leisure hours played the viola with a string quartet, regularly practicing in the dining room, where the acoustics was best. And he came home each day at noon expecting lunch to be served, and afterwards dozed off at the canapé; flakes of cigar ash occasionally dripping at his well-filled waistcoat. She loved him.

In trips down memory lane she recalled their early days together in the 1890s, when he still had his High Wheeler without brakes, on which he in early days had inspected the harbour high in the saddle, riding in a haughty style as if head and shoulders above the workers—for which his oldest children was to reproach him—but creating amusement among the folks by reeling at the brink of the dock, and more than once avoiding a dip in the water by running into a tree or a shipload of timber; 'what a waste of talent!'

With a whole clan flocking to their abode, passing in and out at all hours, Emmy threw tea parties with drop-in guests and musical entertainment each and every day—their offspring remembered the noise, high spirits, and laughter at home, and how impossible it was to do any homework there (Figure 35).

Her husband and her father called her 'nimble' and said of her that she was quick at her work. In her youth she was unabashed and a live wire. On the passages from America through New Caledonia and New Guinea to Australia, and the roundabout back to the Pacific Line Islands through New Zealand, she had shown herself forwards in the meeting with indigenous people of Oceania, who reportedly had earlier been cannibals, asking them which part of the body was tastiest. A straightforward question from a rosy marriageable European girl safely guarded by captain, chief officer and interpreting local trade partners, insisting on candid and detailed answers. She was invariably met by the retort that this was a cult ceremony of the past and by no means a feast or a meal, but connected with wars; yet an old man in the South Island harbour conceded that the palm of the hand was the

Fig. 35. *Emmy with her family in the Central Palace, Gefle, c.1910.* Courtesy of the Maria L. Hallengren Collection.

tastiest part. Smoked, as frog's legs maybe, she considered. The adolescent had wondered which were the most delicious parts of the female body. Thighs had been her (and the crew's) guesstimate.

Her sister Maria criticized her for actively attending all parties, and dancing all dances with strangers through the nights, whereas she refrained from all such entertainments. When she was asked to dance by the commander at the ball with the Russian navy, she had transferred the invitation to her willing sister, who took the turn round the floor. Mia was the thinker, reader and writer to whose reflective and literary character we owe the knowledge of their trip, and who is the true begetter of the present book and its life story, the source of inspiration for later generations. She spent most time alone in the stern studying and thinking, while her more easy-going and light-hearted sister, who did not care much for books or religion and was no intellectual, assisted the young unmarried cook Johan Axel Waltner (b. 1864, 22 years), sang with the crew or played the piano on board.

By his Gefle string quartet her husband was celebrated on his 75th birthday with a magnificent photograph of the chamber orchestra with their autographs and including a score with the introducing beats of Franz Schubert's string quartet No. 14 in D minor, D. 810, commonly called *Death and the Maiden* (*Der Tod und das Mädchen*) (Figure 36). A piece written by the Austrian composer when he was ill and realized that death was impending. That was the kind of music played in the family.

The musical theme or Leitmotif through the generations stem from Grandmother Amalia Mathilda, the girl who died at 24—leaving with her two daughters a piano in her estate of 1873, in the deed then valued at 20 kr —the very square grand piano under which they all had spent the night of horror in the city conflagration of 1869, covered by blankets. She played it, and it is known that she played well.

At the age of 79, the viola player Gotthard, having performed the alto part for years, fell on the hall floor and lay senseless. His upset wife and young maid hurried to help him and tried to resuscitate him, but he remained lying paralyzed at their feet, and when he finally opened his eyes among the nervously rustling skirts surrounding him and looked up along the white silk stockings he just said 'What beautiful legs you have!'

Fig. 36. *Emmy's home became a centre for musical life. Her husband with his Gefle String Quartet in 1932.* Photograph owned by the author.

A long life in the service of trade and shipping thus ended in 1936, but his wife and six children survived him. These progenies—Greta, Sven, Bo, Stig, Gunborg, Ingrid—were all present when his wife died at 76. Emmy was kept alive for the last eight years by her cherished family but wanted to die from a stroke like her husband. When it finally struck her and she woke up at the hospital, she cried: '*Det var en strunthjärnblödning!*' ('That apoplexy was poor show!') and emptied the bedside carafe at a draught. But her condition soon changed for the worse.

Three days after midsummer of 1944, the Blessed Sacrament was given to her and all children on the deathbed, officiated by the Gefle curate Alvar Cedermark (b. 1904) in holy vestments. The clergyman afterwards said that this last supper with the six profoundly touched children assembled around their beloved expiring mother was the most impressive moment in his life. He lived to reach a great age and remained a family priest for half a century.

It is said that all is in a state of flux, but it is rather in circulation in forwards movement. 'All streams flow into the sea, yet the sea is never full. To the place the streams come from, there they return again'. The 1.3 billion cubic kilometres of water on earth has not changed for a billion years as it forms a closed system [Green, 1995]. In this book we have seen a continuity in all that came about, and the oceans as the backdrop and connecting link that have joined so many cultures and life stories to one another and determined family callings and careers. And overall, in this case the plot had a happy ending, due to the headway made.

The American philosopher Ralph Waldo Emerson, once studied by Maria Söderström, in 'Spiritual Laws' compared the course of the individual to that of a ship. 'He is like a ship in a river; he runs against obstructions on every side but one, on that side all obstruction is taken away and he sweeps serenely over a deepening channel into an infinite sea' (*Essays: first series*, 1841).

VII. Voyage to Unexplored Regions of the Earth

1. Family Business

THE FACT THAT MIA'S DAUGHTER GURLI, NAMED after her father's ship, was married into the enterprising Herrström family of Scania, and her sister Emy espoused an official in Gefle harbour, Gotthard Hallengren, the son of the councillor who had organized rescue operation at the city conflagration of 1869, had remarkable maritime consequences and meant a continuation of the nautical history of the family. A strange and illustrative story is formed by the following series of events, which had a notable outcome in the voyages of discovery, and the naming of uncharted lands.

Gotthard's brother Johan Jacob Hallengren, an accountant and commercial agent born in 1852, moved from his hometown of Gefle to Stockholm, and settled down in the Wahrenberg district on Norrmalm in the St. Jacob and St. John parish in 1890.

In the beginning of the decade, and together with the building proprietor Oscar Herrström (b. 1857), he founded the well-stocked grocery store and market hall *Hallengren & Herrström*, which at Klarabergsgatan 25 in Stockholm, with national telephone number 3134, took up the competition with the famous upper-class grocer's shop Arvid Nordquist at Sturegatan 38 and thus launched a modern import company and supermarket ahead of its time.

The firm's magnificent forty-three-page *Price-courant* of spices, preserves, delicacies, and wines from 1897 shows the wide range of the assortment (Figure 37). The wine list includes a large group of Bordeaux and other French wines, which Maria Söderström's great-grandchild, the artist Pontus Carle Herrström (b. 1955), living in Bordeaux, today recognizes as château wines produced by his life partner, the actress and

Fig. 37. *The Hallengren & Herrström catalogue of 1897, printed by Victor Petterson, Stockholm 1897.* Copy in the Royal Library, the National Library of Sweden.

singer Gwénola de Luze's Franco-Swiss family and friends—including Grand Vin Château Ducru-Beaucaillou 1872 (4 kr) from Nathaniel Johnston & Fils. The price list also includes other splendid beverages, with prices set in Swedish kronor and öre.

In comparison it should be observed, that the annual income in the fairly well-to-do metropolitan middle class rarely exceeded 3,000 kr, so the selection is exclusive. Johan's brother Gotthard, the harbour clerk in Gefle, earned 900 kr a year. You need to multiply by 100 to arrive at a somewhat comparable prize level today.

Champagne wines, *large bottles*.
G. H. Mumm & C:o, Cabinet Crémant 6:75
Heidsiek & C:o, Monopole 7:75
" Grand Vin Royal 6:75
Charles " Grand Vin Médaillé 6:75
Grand Crémant (Minet Jeune) Dufaut & C:o, Château de Pierry 5:50
Grand Vin Métropole, *recommended as excellent* 5 –
Grand Vin l'Exposition Sillery, Marquis de Marillac 4:25
Grand Duc (Carte Blanche) 3:50
La Perle 3 –
Cascade 2:50

The background is as follows. In 1880, Oscar Herrström had married the Gefle girl Eva Oest (1857–1942), a friend of Emmy and especially of her sister Maria. Their eldest son, Torsten Herrström, born the following year, in 1905 married Sully Rabén, daughter of the prosperous wine merchant Karl Rabén (1853–1903). Pontus Carle's father, Stig Herrström in Chinon, grandchild of Mia, has put forward the idea that Oscar through his wife had acquaintances in Gefle, i.e., also with a Hallengren, who was also a businessman, and that his family thereby came into contact with wine merchants, which later also led to the son's marriage to a wine merchant's daughter! This may have mattered, and contacts between the families can be traced further back in the nineteenth century.

In any case, the Hallengren & Herrström Universal Imports became a considerably rewarding but rather short-lived story, and hardly took up the men's time to any great extent. After the turn of the century, the salesman and importer Hallengren disappeared from his lodgings in the Old Town to an unknown place abroad, probably pretty well-heeled. Herrström diligently built up the fortune as a building contractor in Stockholm, where he laid his hands on the Grand Hotel Winter Garden, the Strand Hotel, the Johnson Line House at Stureplan; Grand Hotel in Saltsjöbaden, and Sundsvall's Enskilda Bank's palace on the fabulous Fredsgatan (Peace Street) No. 4.

A particularly significant and historically lasting result of the private capital that was created, however, was Herrström's sponsorship of Otto Nordenskjöld's research expedition with the ship *Antarctic* 1901–1904, where previously unknown areas around the South Pole were visited, explored, mapped, and named.[1]

Over time, new geographical place names in Antarctica in this way appeared, such as the *Prince Gustav Channel*, 64° 10′S 57° 45′W; and *Herrström Island* in the Weddell Sea—also known as the *Herrstrom Islet*; or *Isla Herrstrom*, Chile—in the archipelago around James Ross Island at 63° 50′0″ S 58° 15′0″ W, which was separated from the vast continent's protruding peninsula, the *Antarctic Peninsula*, by the sound that *Otto Nordenskjöld* named after the Swedish Crown Prince in 1903.

This is clearly documented in *South America Pilot*, II (United States Navy Hydrographic Office: Washington GPO, 3rd ed. 1929; Government Printing Office, 1930), as well as by *Sailing Directions for Antarctica: Including the off-lying Islands South of Latitude 60° S* (Hydrographic Office Pub. No. 27, US Government Printing Office, Washington DC, 2nd ed. 1960 [1943]), and moreover in *Diccionario de nombres geográficos de la costa de Chile*, Vol. 3 (Instituto Hidrográfico de la Armada, Santiago 1974); and in Andrew Taylor's famous *Two Years Below The Horn: Operation Tabarin, Field Science, and Antarctic* (University of Manitoba, 2017).[2]

---

[1] Nordenskjöld, 1977.
[2] See further the full account and cartographic material in Otto Nordenskjöld's *Antarctic: två år bland Sydpolens isar*, I-II, Stockholm: Bonnier 1904, where the Herrström Island is found in the maps.

2. A Coincidental Encounter

Numerous places on the globe have been named after the explorer himself. On the world maps we find the *Nordenskjöld Coast, Nordenskjöld Glacier, Nordenskjöld Peak, etc.*...

After decades of exploring the Patagonian ice fields and Antarctica, a peak in his long career, the world-famous polar explorer and geographer Otto Nordenskjöld (b. 1869 and of the same age as the sisters Emy and Mia, the female circumnavigators of the world) finally returned to Gothenburg in Sweden, and in 1920 it was announced that he was to lead a course in geography at the university there. Finally, the unique opportunity was given to have the world-famous explorer as a teacher!

However, only one student showed up at the roll call!

It was Emy's son Sven, the grandchild of Captain Söderström.

The one-student-audience did not make the professor call off the one-member university course but turned it into a series of intense private lessons, followed by long periods of practice. The student was sent off on expeditions to learn the art of land surveying by triangulation and the technique of measurement.

And from the outset he was thoroughly examined in the basics of geography. It all started with oral examination of definitions.

QUESTION: What is climate?

ANSWER: Temperature and precipitation!

The exam book is still extant (Figure 38).

A few years later, Nordenskjöld, the world traveller and discoverer of the vast expanses of ice and distant islets, was hit by a bus in Gothenburg on his way home and died on 2 June 1928. In his pocket he had a ticket for a new long journey, and had no money for a return ticket.

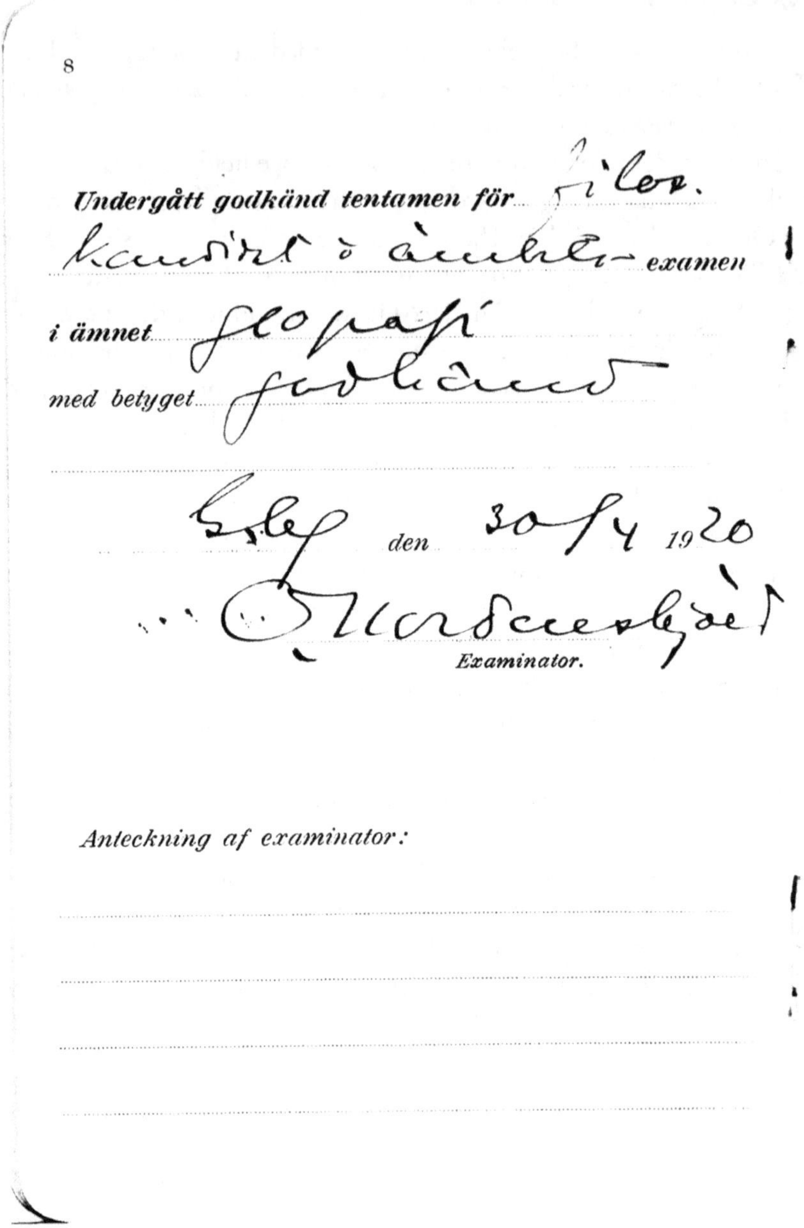

8

Undergått godkänd tentamen för filos. kandidat o. ämbets- *examen*

i ämnet geografi

med betyget godkänd

Gbg *den* 30/4 *19*20

O. Nordenskjöld

Examinator.

Anteckning af examinator:

Fig. 38. *The polar explorer Otto Nordenskjöld passed his only student in geography, Emmy's son*. The certificate in the examination book of Sven Hallengren from Göteborgs Högskola, the original owned by the author. Author's photograph.

Bibliography: Sources and Literature

Ahlström, Gunnar. *Det moderna genombrottet i Nordens litteratur*. Stockholm: Rabén & Sjögren, 1973 (1947).
American Library Association (ALA): 'Indigenous Tribes of Seattle and Washington', November 19, 2018, listed 29 tribes (n/a). Identical information was made available by the Association of College & Research Libraries (ACLR) conference website in 2021: <https://careers.uw.edu/blog/2021/09/10/the-indigenous-tribes-of-seattle-and-washington-a-guide-for-employer-and-community-understanding-created-by-acrl-2021/>. Updates are provided by the American Indian Library Association (AILA), <https://ailanet.org>. The Washington State Department of Social and Health Services (Tacoma, WA) <https://www.dshs.wa.gov> has likewise listed twenty-nine tribes in Washington State, whereas the Governor's Office of Indian Affairs (Olympia, WA) in fact lists thirty-five Indian tribes and settlements including out-of-the-state locations of some of them: <https://goia.wa.gov/tribal-directory/federally-recognized-indian-tribes>.
Archives and collections of Sjöhistoriska museet, the Maritime Museum of Stockholm: Statens maritima och transporthistoriska museer, Stockholm, Sweden, https://www.smtm.se (SMTM) Box 27131, 102 52 Stockholm, Sweden. Sjöhistoriska museet, Djurgårdsbrunnsvägen 24, Stockholm; <https://www.sjohistoriska.se>.
Archives, Gävle sjömanshus (MSS), the Gefle mercantile marine office, now in the Swedish National Archives. Riksarkivet, Stockholm, Sweden, <https://riksarkivet.se>. <https://sok.riksarkivet.se/?postid=ArkisRef+SE%2FHLA%2F1170001> 'Sjömanshuset i Gävle (1754 – 1969)'. Landsarkivet i Härnösand (depå: Kusthöjden 46).
Archives of Gothenburg mercantile marine office (MSS); Sjömanshuset i Göteborg. Riksarkivet, the Swedish National archives. Riksarkivet, Stockholm, Sweden. <https://riksarkivet.se>. 'Sjömanshuset i Göteborg 1753-1967' (Landsarkivet i Göteborg, depå: Arkivgatan) <https://sok.riksarkivet.se/arkiv/wT4w3gHtrH6cyG018W43t3>.
Archives of Kommerskollegium (MSS), the Swedish College of Commerce, in the National Archives. Riksarkivet, Stockholm, Sweden. <https://riksarkivet.se>. 'Kungliga kommerskollegiums arkiv (1651–)' <https://sok.riksarkivet.se/arkiv/cImhhecRrH6d0G02H087k3> Riksarkivet, Box 7223, 18713 Täby, Sweden.

Archives, Sjömanshusets i Söderhamn arkiv (MSS), the Swedish National Archives. Riksarkivet, Stockholm, <https://riksarkivet.se>. Archives, 'Sjömanshusets i Söderhamn arkiv (1818-1962)'. Landsarkivet i Härnösand (depå: Kusthöjden 46) <https://sok.riksarkivet.se/arkiv/cgY7ruzo02PTbnoZmsXp57>.

Archives, Västerviks sjömanshus (MSS), the mercantile marine office, now in the Swedish National Archives. Riksarkivet, Stockholm, Sweden, <https://riksarkivet.se>. 'Sjömanshusets i Västervik arkiv (1753-1961)'. Landsarkivet i Vadstena (depå: Vadstena Castle) <https://sok.riksarkivet.se/arkiv/Zf7vyavqQWZTuu1eMvyUeB>.

Aspegren, Ebbe. 'De första svenska världsomseglingarna'. In: *Forum navale*. Stockholm: Sjöhistoriska samfundet, ISSN 0280-6215; 1982 (37), pp. 1–72; illustrated.

Atmer, Ann Katrin. *100 år i lägenhet: svensk stadsbostad och närmiljö 1870 till idag* [The Swedish flat: urban housing in Sweden 1870 to today]. Stockholm: Sveriges arkitekturmuseum, 1976.

Baker, Robin (ed.). *The Mystery of Migration*. London: Macdonald, 1980.

Baker, William A., Sam Svensson & George P.B. Naish. *The Lore of Sail*. New York: Facts on File, 1984.

Baudelaire, Charles. 'L'Albatros'. In: *Les fleurs du mal: édition de 1861*, by Charles Baudelaire. Paris: Gallimard, 1975. A critical edition introduced, established, and annotated by Claude Pichois.

Beijbom, Ulf. *Australienfararna*. Stockholm: LT, 1983. [An illustrative account of Swedish emigration on the Australian trade route, and how so many Swedish sailors jumped ship and disappeared in the harbours, in the 1880s sometimes one per day. Ships like the *Gurli* reported missing men; many were not reported at all, as the captains did not miss them much. The agricultural crisis in Sweden and attractive fields and raw material resources were the driving forces, here as well as in the mass emigration to America. The gold rushes of the middle of the century had set it all in motion.]

Bell, Christopher M. *Churchill and the Dardanelles: Myth, Memory, and Reputation*. Oxford: Oxford University Press, 2020.

Berge, Lars. *Monstervågen: en studie av sanningshalten i matros J.W. Granströms äventyr på de sju haven 1914–1915*. Stockholm: Albert Bonniers Förlag, 2020.

Birkeland, Torger & Joshua Green. *Echoes of Puget Sound: Fifty Years of Logging and Steamboating*. Chicago: Papamoa Press, 2019 (©1960).

Björk, Arvid (ed.). *Skepparesocieteten, nuvarande Sjökaptensföreningen, i Gefle [1776–1925]: Bidrag till dess historia, samlade och utgivna av Styrelsen*. Gefle: Sjökaptensföreningen, 1926.

Blomberg, J. Gustaf. *En svensk sjökaptens sanna upplevelser och äventyr som man, styrman och befälhavare å segelfartyg samt tvenne år som befälhavare å ångare trafikerande La-Plata floden och brasilianska kusten*. Stockholm: Seelig, 1934. [This memoir includes an account of Blomberg's experiences as chief mate of the *Atlantic* under the command of captain Johan Axel Söderström.]

Bly, Nellie (Elizabeth Cochrane). *Around the World in Seventy-Two Days*. Rockville, Maryland: Wildside Press, 2009 (1890). [An account of her trip round the world in 1889–1890 by various means of transport.]

Bolster, W. Jeffrey. *The Mortal Sea: Fishing the Atlantic in the Age of Sail*. Cambridge: Belknap Press of Harvard University Press, 2012.

Bougainville, Louis Antoine de. *Voyage de Bougainville autour du monde ... raconté par lui-même*. Paris: Bibliothèque d'aventures et de voyages, 1924.

Bowditch, Nathaniel. *The New American Practical Navigator: Being an Epitome of Navigation; Containing all the Tables Necessary to be Used With the Nautical Almanac, in Determining the Latitude and the Longitude by Lunar Observations and Keeping a Complete Reckoning at Sea*. New York: E.& G.W. Blunt, 1836.

Brassie, Annie. *A Voyage in the 'Sunbeam': Our Home on the Ocean for Eleven Months*. Chicago: Belford, Clarke & Co.; Donohue & Henneberry, Printers and Binders, 1881.

Bremer, Fredrika. *Hemmen i den nya verlden: en dagbok i bref, skrifna under tvenne års resor i Norra Amerika och på Cuba*, 3 vols. Stockholm: Norstedt, 1853–1854. [Travel writer esteemed in the Själander school circles.]

Bremmer, Fatima. *Ett jävla solsken: en biografi om Ester Blenda Nordström*. Stockholm: Forum, 2017.

Brewster, Mary & Joan Druett. *She Was a Sister Sailor: The Whaling Journals of Mary Brewster, 1845–51*, ed. Joan Druett. Mystic, Connecticut, USA: Mystic Seaport Museum, 1992. American Maritime Library, vol. 13.

Brooke, Michael: *Far From Land: The Mysterious Lives of Seabirds*. Princeton: Princeton University Press, 2018.

Brown, Nancy Marie. *The Far Traveler: Voyages of a Viking Woman*. Orlando: Harvest Books Harcourt, 2008. ['Five hundred years before Columbus, a Viking woman named Gudrid Thorbjarnardóttir sailed off the edge of the known world.']

Cagner, Ewert, Bengt Kihlberg & Sam Svensson. *L'arte navale: enciclopedia nautica illustrate*. Milan: Mursia & Co., 1964.

Channing, William Ellery. *Ett fullkomligt lif menniskans mål: tolf predikningar [= The Perfect Life]*, trans. Maria Söderberg. Upsala: R. Almqvist & J. Wiksell, 1881.

Chapman, Heather, and Judith Stillman. *Melbourne Then & Now*. Wingfield, South Australia: Cameron House, 2007.

Charlesen, Janne. 'Linjedop i dåtid och nutid'. In: *Blekinge ångbåt & segel*, 1997, pp. 17–26. *Blekinge ångbåt & segel* is an annual maritime publication published by Blekinge sjöfartsmuseum in Karlshamn, Sweden.

Chatterton, E. Keble. *Dardanelles Dilemma: The Story of the Naval Operations ... With 69 Illustrations and 10 Maps*. London: Rich & Cowan, 1935.

Chatterton, E. Keble. *Sailing Ships: The Story of Their Development From Earliest Times to the Present Day*. London: Sidgwick & Jackson, 1914.

Cohen, Margaret. *The Novel and the Sea*. Princeton and Oxford: Princeton University Press, 2010.

Coleridge, Samuel Taylor & Paul H. Fry (ed.). *The Rime of the Ancient Mariner: Complete, Authoritative Texts of the 1798 and 1817 Versions With Biographical and Historical Contexts* ... Boston: Bedford/St. Martin's, 1999.

Conrad, Joseph. *The Mirror and the Sea*. London; New York: Methuen & Co; Harpers, 1906. Memoirs of Józef Teodor Konrad Korzeniowski's voyages in the merchant marine, 1874–1893.

Cook, James & Philip Edwards. *The Journals of Captain Cook*. London: Penguin, 2003.

Cookery for Seamen. Greenwich, London: Natural Maritime Museum, 2019 (1894). Facsimile edition of Cookery For Seamen by Alexander Quinland and N. E. Mann, published in 1894 by Tinlin and Co. Printers, Victoria Street, Liverpool (Price: sixpence), 75 pages. With a new introduction by Stawell Heard, Librarian of the National Maritime Museum.

Cordingly, David & John Falconer. *Pirates: Fact & Fiction*. London: Collins & Brown, 1992.

Couper, Alastair (ed.). *The Times Atlas of the Oceans*. London: Times books, 1983.

Cox, Edward Godfrey. *A Reference Guide to the Literature of Travel: Including Voyages, Geographical Description, Adventures, Shipwrecks and Expeditions*. 3 vols. Seattle: University of Washington Press, 1935-1949. Series: University of Washington publications in language and literature, 0085-7947.

Dana, Richard Henry Jr. *Two Years Before the Mast*. New York: Harper and Brothers, 1840.

Danielsson, Bengt. *Gauguin in the South Seas*, trans. Reginald Spink. London: G. Allen & Unwin, 1965.

Danielsson, Bengt. *Love in the South Seas*. New York: Reynal, 1965.

Danielsson, Bengt. *What Happened on the Bounty*. London: G. Allen & Unwin, 1962.

Darwin, Charles. *The Voyage of the Beagle,* ed. Janet Browne and Michael Neve. London: Penguin Books, 1989.

Davidsson, Jan. *Mentor Ohoj: barkskeppet med hemort i Mollösund, Göteborg, Råå 1876–1917*. Lund: Signum, 1981.

De Pauw, Linda Grant. *Seafaring Women*. Pasadena: Peacock Press, 1998.

Diccionario de nombres geográficos de la costa de Chile, vol. 3. Santiago: Instituto Hidrográfico de la Armada [the Chilean Navy, the naval warfare service branch of the Chilean Armed Forces], 1974.

Diston, John, Murdo Downie & Alexander Ingram. *The New Seaman's Guide: and Coaster's Companion, Containing Complete Sailing Directions for Ships ... and All Useful Tables, viz. Tables of the Magnetic Courses, and Distances From Place to Place, for all the Coasts of Europe, and Many of Africa and Asia ... Copious Tables of Latitudes and Longitudes* ... Edinburgh: D. Schaw and Son, 1800.

Drake, Francis. *The World Encompassed*. New York: Da capo Press, 1969. (Facsimile; London 1628). *Series*: The English Experience; 103.

Dufour, Auguste-Henri. *Océanie*, Dressée par A. H. Dufour, Gravée par Charles Dyonnet. Paris: Armand Gilbert Le Chavalier éditeur, Rue Richelieu 61, 1863.

[Comprehensive overview of the Pacific Ocean consulted by the officers of the *Atlantic* before studying nautical charts. With detailed maps of southern Australia, New Zealand, New Caledonia, the Marquesas Islands, Gambier Islands, and Tahiti. Malden Island sits almost in the middle of the map. Cartographic material owned by the author (Anders Hallengren).]

Duyker, Edward. *Nature's Argonaut: Daniel Solander 1733–1782: Naturalist and Voyager With Cook and Banks.* Melbourne: Miegunyah Press, 1998.

Edman, Stefan & Philip Plisson. *Planeten havet* [text: Stefan Edman; photo: Philip Plisson]. Stockholm: Max Ström, 2010.

Ekelund, John. *Gefle stads hamn: redogörelse för anläggningarne vid Fredriksskans jämte återblick på hamnens historia.* Gefle: Serranders tryckeri, 1905.

Ekman, Henrik. *Enskilda barnhemmet Salem: 1859–1935: Historik.* Gävle: Lantmännens tr., Westlund & Co., 1937.

Emerson, R. W. 'Introductory lecture on the times'. In: *Miscellanies by Ralph Waldo Emerson*, by John Morley. London: Macmillan, 1884. Series: The works of Ralph Waldo Emerson, vol. I.

Emerson, R.W. *The Essays of Ralph Waldo Emerson* [Text established by Alfred R. Ferguson and Jean Ferguson Carr. Introduction by Alfred Kazin]. Cambridge: Belknap Press, 1987.

Engvall, Margareta. *Staden Brinner!* Hedesunda: Knight, 2008.

FreeHand, Julianna, Alice Drinkwater & Sumner Drinkwater. *The Only Woman on Board: The Photographs, Diaries, Letters, and Memorabilia of a Maine Sea Captain and his Wife, 1859–1908.* Camden: Picton Press, 1994.

Freycinet, Rose de. *A Woman of Courage: The Journal of Rose de Freycinet on her Voyage Around the World, 1817–1820*, trans. Marc Serge Rivière. Canberra: National Library of Australia, 2003 (1996).

Garberg, Rickard Wilhelm & Sven Garberg. *Ur kapten Garbergs "loggbok": anteckningar och minnen / sammanställda av Sven Garberg.* Gävle: Hallbergs bokhandel, 1962.

Garberg, Rickard Wilhelm. *I rigg och skans: minnen från en djupvattenseglares hundår.* Stockholm: Wahlström & Widstrand, 1936. [This memoir contains encounters in London and Kingston with captain Axel Söderström and the deckhand Pehr Johan Zellinger (here named 'John Zellund') of the full-rigger *Thor* in 1870–71.]

Goetzfridt, Nicholas J. *Indigenous Navigation and Voyaging in the Pacific: A Reference Guide.* Westport: Greenwood Press, 1992. Series: Bibliographies and indexes in anthropology; 6.

Grangert, Lennart. *Glimtar ur Gävle stadsfullmäktiges historia 1863–1963.* Gävle: stadsfullmäktige, 1963. [About Jacob Hallengren (1822–1878) and other members of the Gefle city council.]

Green, Bill. *Water, Ice & Stone: Science and Memory on the Antarctic Lakes.* New York: Harmony Books, 1995.

Gurlitt, Cornelius. *C. Gurlitt's Etudeverk för piano*, I–III, Op. 50–52. Stockholm: Huss & Beer, n.d. [Scores owned and studied by the cinema pianist Emmy Söderström.]

Hägerhäll, Bertil. 'Släkten Brodin'. *Båtologen:* medlemsblad för Klubb Maritim, förening för fartygshistorisk forskning, Vol. 59 (2021), No. 6, pp. 362–385. Viken: Klubb Maritim (Sweden), 2021.

Hägg, Erik & Jacob Hägg. *Under segel: en skildring från segelflottornas tid.* Stockholm: Lindfors, 1935.

Hallengren, Anders. *Campagna per la felicità: l'avventura caprese e napoletana di Anne Charlotte Leffler, duchessa di Caianello.* San Michele: Anacapri, 2001. [On the modern breakthrough of woman authors and travellers.]

Hallengren, Anders. *The Moment Is Now: Carl Bernhard Wadström's Revolutionary Voice on Human Trafficking and the Abolition of the African Slave Trade,* ed. Anders Hallengren. West Chester: Swedenborg Foundation, 2019. [On the dark aspects of utilizing the trade winds in the Age of Sail.]

Hallengren, Anders. 'Transporter på haven' [On maritime transport]. *Värld och Vetande,* 1982; 31(1). Published in Västra Frölunda, City of Gothenburg: 1982. [Due to the cost efficiency of the commercial tonnage measured in cost per ton/kilometres (tkm), shipping remains the cheapest way of conveying cargos long distances. It is furthermore environmentally friendly, and compatible with sustainable development.]

Hallengren & Herrström. *Pris-Courant à specerier, konserver, delikatesser och viner.* Stockholm: Victor Petterson, 1897, 43 pages. OCLC Number 186020860. Copy in the Royal Library (Kungliga Biblioteket), Stockholm, Sweden. A digital version produced by the library in 2012 was made available at: <http://libris.kb.se/bib/14800303>.

Haller, Stephen A. *Families at Sea: An Examination of the Rich Lore of "Lady Ships" ... 1850–1900.* San Francisco: National Maritime Museum Association, 1985.

Hamilton, Douglas & John McAleer (eds.). *Islands and the British Empire in the Age of Sail.* Oxford University Press, 2021.

Hancock, Sir William Keith. *Politics in Pitcairn and Other Essays.* London: Macmillan, 1947.

Helmerson, Klas. *Segelfartyg: skepp till havs och på brev.* Stockholm: Posten Frimärken, 2008. Text in Swedish and English.

Henricson, Ingvar. '1909, "Atlantic"—sista överhalningen.' *Gefle Dagblad,* 18 July, 1997. [On the last overhaul of the Atlantic in the dock of Brodin's wharf in the Gavle River in 1909.]

Henricson, Ingvar. 'Atlantic—Gävles sista djupvattenseglare.' *Gefle Dagblad,* 14 May, 1983.

Henricson, Ingvar. *Inte bara en skepparhistoria.* Hudiksvall: Winberg, 1990.

Henricson, Ingvar. 'Mia och Emy på världshaven.' *Gefle Dagblad,* 3 Augusti, 2007.

Henricson, Ingvar. *"Mister Agent, We Are Sinking!": öden och äventyr kring Haegerstrands, på världshaven sen 1859.* Göteborg: Breakwater Publishing, 2004.

Henricson, Ingvar & Tom Sandstedt. *Drömmar om seglande skepp: Gävles sjöhistoria i kaptenstavlor.* Gävle: Länsmuseet Gävleborg, 2003.

Herrström, Stig. 'Kommentarer till Mias dagbok' (1997), notes updated in 2009. Dossier in manuscript. The Hallengren Collection (MSS). Private possession.

Herrström, Yngve. Genealogical notes (MSS) of various age, 1973 (updated 1979–1999). Copy in the Hallengren Collection. Private possession.

Hoffman, Michael. *Messing About in Boats*. The Clarendon Lectures. Oxford: Oxford University Press, 2021.

Höglund, Patrik. *Skeppssamhället: Rang, roller och status på örlogsskepp under 1600-talet*. Lund: Universus Academic Press, 2021. Series: Södertörn doctoral dissertations, 1652-7399; 195. Series: Forum navales skriftserie, 0280-6215; 77. Thesis. Huddinge: Södertörn University, 2021. With an English summary. [In seventeenth century Sweden, during the period as a great maritime power, there were no women in the crews, but some worked aboard the ships disguised as men, and in times of peace officers were permitted to bring wives and children on Swedish warships.]

Horsburgh, James & Edward Dunsterville. *The India Directory, or: Directions for Sailing to and from the East Indies, China, Japan, Australia, and the Interjacent Port of Africa and South America*. London: Wm. H. Allen, 1864.

Humbla, Philibert, Nils Norling, and G.E. Lundén, eds. *Ur Gävle stads historia: utgiven till femhundraårsjubileet*. Gävle: Lantmännens tryckeri Westlund & co., 1946. [A standard work on the history of Gävle City with a number of contributing specialists; published on behalf of the city council. 851 pages.]

Illum, Peter & Ole Mortensøn. *Skipper Peter Illums dagbog 1804–1893*. Faaborg: Faaborg Byhistoriske Arkiv, 1992.

Imray, James Frederick. *Chart of the Indian Ocean, Showing the Whole Navigation Between the Cape of Good Hope and China, Australia, New Zealand &c, ...* London: James Imray & Son, 1853. [Nautical chart owned by captain Axel Söderström, the *Atlantic*.]

Imray, James Frederick. *Chart of the West, South and East Coasts of Australia Extending From the Houtman Abrolhos Rocks to Moreton Bay and Including the Island of Tasmania*. London: James Imray and Son, 1853–1863. [Map owned by captain Axel Söderström, the *Atlantic*.]

Imray, James Frederick. *Sailing Directions for the West Coast of North America: Embracing the Coasts of Central America, California, Oregon, Fuca strait, Puget Sound ...Containing Various Remarks on the Winds, Tides, Currents*. London: James Imray, 1853. [Nautical chart owned by captain Axel Söderström, the *Atlantic*.]

Imray, James Frederick. *South and East Coasts of Australia* [in four charts]. Chart No. 1 Australian Bight to Cape Northumberland. Compiled by James F. Imray. London: James Imray & Son, 1872 (1867). [Copy owned by the captain of the *Atlantic*.]

Irving, R.A. & T.P. Dawson. *The Marine Environment of the Pitcairn Islands. A Report to Global Ocean Legacy, a Project of the Pew Environment Group*. Dundee: Dundee University Press, 2012.

Karlsson, Lina, Susanne Sellin & Sanne Vader. *Bostadens historia och framtida utveckling [The history and future development of residential housing]*. Växjö: Institutionen för Teknik, Linnéuniversitetet, 2012.

Kennedy, Gavin. *Captain Bligh: The Man and his Mutinies*. London: Duckworth, 1989.

Kent, Neil. *The Soul of the North: A Social, Architectural and Cultural History of the Nordic Countries, 1700–1940*. London: Reaktion Books, 2001. Series: Histories, cultures, contexts.

Kjellgren, Sven. *Gamla gårdar och släkter i 1800-talets Westervik*. Stockholm: Impactum, 2014.

Kjellgren, Sven. 'Gamla tiders Tjust: Svartsjukedrama till havs'. *Dagens Västervik*, 15 September 2020.

Krusenstern, Adam Johann von [Kruzenštern, Ivan Fedorovič]. *Atlas de l'Océan Pacifique dressé par ... Commodore de la Marine Impériale de Russie*. Saint-Pétersbourg: L'Imprimérie du Département de l'instruction publique, 1824. (Copy in Krigsarkivet, Stockholm.)

Krusenstern, Adam Johann von [Kruzenštern, Ivan Fedorovič]. *Putešestvie vokrug sveta v 1803, 1804, 1805 i 1806 godah na karabljah 'Nadežde' i 'Neve'*. Moscow: Gosudarstvennoe izdatel'stvo geografičeskoj literatury, 1950. [The Admiral's enlightening account of his circumnavigation with the ships Nadezhda ('Hope', formerly the British slave ship *Leander*) and Neva in the early 1800s. Comprehensive and crammed with facts. Translated into many languages.]

Lagerqvist, Lars O. *Vad kostade det?—priser och löner från medeltid till våra dagar* . 7th ed. Lund: Historiska Media & The Royal Coin Cabinet, 2015.

Larsson Hallengren, Maria [usually called Maria L. Hallengren]. 'Släktforskning: Min farfar Bo Hilding Hallengren.' Accessed October 13, 2021. <http://www.hallengren.se/maria/slaktforskning/Farfar/index.html> --a genealogy created by a grandchild of the shipbroker Bo Hallengren and great-granddaughter of the circumnavigator Emmy Söderström.

Leland, Lilian. *Traveling alone: A woman's journey around the world*. New York: From the Press of John Polhemus; Trade Supplied by The American News Company, 1890. [An interesting account compiled from letters.]

Leslie, Robert C. *A Sea-Painter's Log*. London: Chapman and Hall, 1886.

Lindström, Anders & Gert Malmberg. *Svensk sjöfartshistoria: i storm och stiltje*. 2nd ed. Gothenburg: Breakwater Publishing, 2015.

Linklater, Elizabeth. *A Child Under Sail*. London: Mariners Library 2, 1949. [An account of life on a windjammer].

Ljungström, Jan G. *Glimtar från förr: axplock ur Gävles historia*. Hedesunda: Knight, 2007.

Longitude: Tidskrift från de sju haven: *A Magazine of the Seven Seas* [periodical]. Stockholm: Carlstedt, 1966–1999.

Lundström, Nils Styrbjörn. *Svenska kvinnor i offentlig verksamhet: porträtt och biografier*. Uppsala, 1924.

Lyng, Jens Sorensen. *The Scandinavians in Australia, New Zealand, and the Western Pacific*. Melbourne University Press, 1933.
Macleod, Jenny. *Gallipoli*. Oxford: Oxford University Press, 2015.
Malham, John. *The Naval Gazetteer: or, Seaman's Complete Guide*. London: Allen and West, 1795.
Marmion, R. J. Gibraltar of the South: Defending Victoria: An Analysis of Colonial Defence in Victoria, Australia, 1851–1901. PhD thesis, School of Historical Studies, Faculty of Arts, The University of Melbourne, 2009.
Martin, Bernard. *The Ancient Mariner and the Authentic Narrative*. Melbourne: Heinemann, 1949.
Martin, John Stanley. *Ethnic Identity and Social Organisation in the Scandinavian Community in Melbourne, 1870–1910*. Parkville: University of Melbourne, Department of Germanic Studies, 1982.
Martin, John Stanley (ed.). *Scandinavians in Victoria: Then and Now: The Proceedings of a Symposium at the University of Melbourne to Celebrate the 150th Anniversary of Victoria: Saturday, 3rd August, 1985*. Parkville: Department of Germanic Studies, University of Melbourne, [1987]. *Series*: Monograph series/Scandinavian-Australian Migration Project, 99-0708037-3; 7.
Mazellier, Philippe (ed.). *Le mémorial Polynesien*, 6 vols. Papeete and Taichung: Hibiscus éditions, 1977–1980.
McGrail, Seán. *Boats of the World: From the Stone Age to Medieval Times*. Oxford: Oxford University Press, 2001.
Melville, Herman. *Moby-Dick; or, the Whale*. New York: Harper & Brothers, 1851.
Melville, Herman. *Omoo: A Narrative of Adventures in the South Seas*. New York: Harper & Brothers, 1847.
Melville, Herman. *Typee: A Peep at Polynesian Life*. New York: Wiley and Putnam, 1846.
Middleton, Dorothy. *Victorian Lady Travellers*. Chicago, Ill: Academy Chicago Publishers, 1993.
Mitzlaff, Eugenia von. *Uppå korsets väg till lifsens krona*. 2 vols. Stockholm: Bonnier, 1879, translated from the German: Durch Kreuz zur Krone (Halle an der Saale: Julius Fricke, 1864). [Read by Maria Söderström during her trip round the world.]
Moorehead, Alan. *Gallipoli*. London: Aurum Press, 2007 [©1956].
Morrison, James. *The Journal of James Morrison, Boatswain's Mate of the Bounty, Describing the Mutiny & Subsequent Misfortunes of the Mutineers, Together With an Account of the Island of Tahiti*. London: The Golden Cockerel Press, 1935.
Murray, Andrew. *Förblifven i Kristus!: Tanken öfver det saliga lifvet i gemenskap med Guds son* [Orig. title: Abiding in Christ]. Stockholm: C. A. V. Lundholm, 1884.
Mutiny on the Bounty: 1789–1989: An International Exhibition to Mark the 200th Anniversary 28 April 1989—1 oct 1989. London: Manorial Research PLC in association with the National Maritime Museum (Greenwich),1989.

Myrdal, Gunnar [Assisted by Sven Bouvin]. *The Cost of Living in Sweden 1830–1930.* London, 1933. Series: Wages, cost of living and national income in Sweden 1860-1930; 1. Series: Stockholm Economic Studies; 2.

Mytinger, Caroline. *New Guinea Headhunt.* New York: The Macmillan Company, 1946.

Natt och Dag, Svante & Richard Melander (ed.). *Jorden rundt under svensk örlogsflagg: ögonblicksbilder från fregatten Vanadis' verldsomsegling 1883–1885* (Ottilia Adelborg illustrator). Stockholm: Bonnier, 1887.

Newby, Eric. *The Last Grain Race.* Boston: Houghton Mifflin, 1956. [On the last Scandinavian ship on the grain trade route, which the Atlantic, the Gurli, and other Swedish ships had travelled].

Nicholson, Ian Hawkins. *Log of Logs: A Catalogue of Logs, Journals, Shipboard Diaries, Letters, and all Forms of Voyage Narratives, 1788 to 1988, for Australia and New Zealand and Surrounding Oceans*, 3 vols. Yaroomba:The Author jointly with the Australian Association for Maritime History, 1990–1998. (Roebuck Society Publication Nos. 41, 47, 52).

Nordenberg, Sven-Olof. *Gävles segelfartyg i utrikes fart sedan 1750-talet.* Hedesunda: Knights förlag, 2014. [The standard work on the merchant navy of Gefle in the age of sail. An indispensable catalogue of ships.]

Nordenskiöld, Adolf Erik [1832–1901]. *Vegas färd kring Asien och Europa: jemte en historisk återblick på föregående resor längs gamla verldens nordkust*, I-II. Stockholm: Beijer, 1880–1881. Trans. *Voyage of the Vega round Asia and Europe* (London, 1881 and New York, 1882).

Nordenskjöld, Otto & Johan Gunnar Andersson. *Antarctica or Two Years Amongst the Ice of the South Pole.* London: Hurst, 1977 [1905]. (Original title: Antarctic: två år bland Sydpolens isar, I-II, Stockholm: Bonnier 1904.)

Nordhoff, Charles & James Norman Hall. *Pitcairn's Island.* Boston: Back Bay Books, 2003 (Little, Brown, and Company, 1934).

Norie, J.W. & W.H. Rosser. *A Complete Epitome of Practical Navigation: And Nautical Astronomy, Containing all Necessary Instructions for Keeping a Ship's Reckoning at Sea: With the Most Approved Methods of Ascertaining the Latitude and Longitude ... Together with a Correct and Extensive set of Tables ...* New ed., considerably augmented and improved. London: Norie and Wilson, 1889 [1877].

Oertling, Thomas J. *Ships' Bilge Pumps: A History of Their Development, 1500–1900.* College Station: Texas A&M University Press, 1996.

Osborne, Richard, John Calambokidis & Eleanor M Dorsey. *A Guide to Marine Mammals of Greater Puget Sound.* Anacortes: Island Publishers, 1988.

Östrand, David. *Släkten Östrand från Hyttön i Älvkarleby.* Stocksund, 1966. [Incorporating minister Carl Östrand's journal of travel 1828 on board the full-rigged ship Victoria of Gefle.]

Parker, Theodore. *Samlade skrifter* [Collected Works], trans. Victor Pfeiff, 10 vols. Upsala: Edquist, 1866–1874. [A philosopher of influence to Mia Söderström.]

Peterson, Matilda & Guy Björklund (ed.). *Matilda Petersons Resejournal: 1872–1874, 1881.* Jakobstad: Jakobstads museum, 1991.
Pettersson, Rune. *Kapten på klipperskepp* [With an introduction by Anders Hallengren]. Stockholm: Bokgillet, 1979.
Pfeiffer, Ida Laura. *A Woman's Journey Round the World.* London: Office of the National Illustrated Library, 1852.
Pitot, Henri. *[Théorie de la Manoeuvre des Vaisseaux réduite en pratique.] The theory of the working of ships, applied to practice. Containing the principles and rules for sailing with the greatest advantage possible,* trans. Edmund Stone. London: C. Davis, and P. Vaillant, 1743.
Rahm, Carl Erik. *Minnen från Australien af en svensk sjöman: med 17 planscher.* Stockholm: Associations-Boktryckeriet, 1880.
Rahm, Carl Erik. *The illustrations in Carl Erik Rahm's 'Minnen från Australien af en svensk sjöman'.* Uppsala: Dahlia Books, 1978.
Rediker, Marcus. *Between the Devil and the Deep Blue Sea: Merchant Seamen, Pirates and the Anglo-American Maritime World.* Cambridge: Cambridge University Press, 1987.
Redogörelse för Högre Flickskolan i Gefle under skolåret 1883–1884. Gefle: Gefle-Posten, 1884. [A school calendar published by the Själander school.]
Ridley, Glynis. *The Discovery of Jeanne Baret: A Story of Science, the High Seas, and the First Woman to Circumnavigate the Globe.* New York: Crown Publishers, 2010.
Ross, Michael Elsohn. *A World of her Own: 24 Amazing Women Explorers and Adventurers.* Chicago: Chicago Review Press, 2014.
Roswall, Fabian Casimir. *Navigation, eller En sjömans dageliga handbok uti styrmanskonsten.* Stockholm: Nestius, 1824. [A time-honoured Swedish text book for steersmen, including how to determine the longitude exactly, written by a former colonel at the Admiralty of Sveaborg.]
Sahlins, Marshall D. *Islands of History.* Chicago: University of Chicago Press, 1985.
Sailing Directions for Antarctica: Including the Off-Lying Islands South of Latitude 60°. 2nd ed. Washington: United States Navy Dept., Hydrographic Office, 1960 (1943). United States Government Printing office.
Sailing Directions for the Baltic Sea From the Sound to the Entrance of the Gulf of Finland: Compiled Chiefly From the Danish, Swedish, Prussian and Russian Surveys. London: Charles Wilson, 1874. OCLC Number 940688131.
Sandemose, Aksel. *A Fugitive Crosses His Tracks.* New York: A. A. Knopf, 1936. [Containing a detailed and authentic description of the Gefle schooner *Ragnar*, built in 1874, and its whole crew, to which the author belonged as deckhand, as well as a scurrilous portrait of its captain Carl August Johansson and his men; 'a phantom ship manned with scum'.]
Schéele, Georg von. *John Groggs minnen från hafven och hamnarne.* Stockholm: C. A. V. Lundholm, 1889.
Schutt, Bill. *Cannibalism: A Perfectly Natural History.* Chapel Hill: Algonquin Books, 2017.

Ship Index. The vessel research database at www.shipindex.org. Published by Peter McCracken. © 2021 Ship Index, LLC. Ithaca, New York 14851, USA.

Sigvallius, Berit. 'Sailing towards the afterlife: Analysis of a ship-formed burial monument at Hellerö by the Baltic Sea'. In: *Dealing with the dead: Archaeological perspectives on prehistoric Scandinavian burial ritual*, by Tore Artelius & Fredrik Svanberg. Stockholm: The Swedish National Heritage Board; 2005, pp. 159–171.

Själander, Karolina. *Karolina Själander och hennes skola: minnesskrift utarbetad till hundraårsdagen av Karolina Själanders födelse 1841 12/9 1941*. Gävle: Hallbergs bokh., 1941.

Skogman, Carl Johan Alfred. [Voyage autour du monde; Erdumsegelung der Königl. schwedischen Fregatte Eugenie.] *Fregatten Eugenies resa omkring Jorden åren 1851–1853*, 2 vols. Stockholm: Bonnier, 1854–1855. [The frigate *Eugenie*, navigated by Christian Adolf Virgin, in 1851–1853 was the first Swedish man-of-war to circumnavigate the globe, a propaganda voyage carried out to promote universal trade and shipping.]

Slocum, Captain Joshua. *Sailing Alone Around the World*. New York: Sheridan House, 1987 (1899).

Smith, Craig & Amanda Demopoulos. 'The Deep Pacific Ocean Floor'. In: *Ecosystems of the Deep Oceans* (Ecosystems of the world), by P.A. Tyler. Amsterdam: Elsevier, 2003.

Snow, Captain Elliot (ed.). *Adventures at Sea in the Great Age of Sail*. New York: Dover Publications, 1986.

Snow, Edward Rowe. *Women of the Sea*. London: Alvin Redman, 1962.

Söderberg, Tom. *Två sekel svensk medelklass: från gustaviansk tid till nutid*. Stockholm: Bonnier, 1972. [Of great relevancy to the subject is the analysis of the middle-class economy of Swedish shipmasters in the mid-1800s, and the precise job description of their work, pp. 225–231.]

Söderblom, Bodil. *Barken Hoppet*. Helsinki: Oy Litorale Ab, 2020. [Monograph about a Finnish barque built in Raumo 1878, and its voyages up to 1910.]

Söderström, Erik. *Mina Gefleminnen: Ett stycke kulturhistoria från 1870-talet*. Stockholm: Esselte, 1940.

Söderström, Maria 'Mia' Mathilda Charlotta. 'Anteckningar under sjöresan' *(Notes During the Sea Voyage)*. Journal dated 7 November 1885–25 September 1887 with additions and supplements. Size c. 21x17 cm, 225 pages (Unpublished MS). Blue cover with a gold rim. Neatly handwritten in Swedish with a pen, a couple of leaves missing. Unruled memorandum-book filled with writing, extant in the family collections. Copy owned by the author. Privat possession.

South America Pilot, II. United States Navy Hydrographic Office: Washington, 3rd. ed., 1929. United States Government Printing Office, 1930.

Sparrman, Anders. [Resa till Goda Hopps-udden, södra pol-kretsen och omkring jordklotet 1772–76, Stockholm 1818] *A Voyage Round the World With Captain James Cook in H.M.S. Resolution*. London: Golden Cockerel Press, 1944. (Transl.

Averil Mackenzie-Grieve; introd. & notes by Owen Rutter; wood-engravings by Peter Barker-Mill).

Stein, Christian Gottfried Daniel & Jakob Melchior Ziegler. *Neuer Atlas der ganzen Erde.* Leipzig: Hinrichs, 1871.

Stein, Werner. *Der grosse Kulturfahrplan: die wichtigsten Daten der Weltgeschichte bis heute in thematischer Übersicht: Politik, Kunst, Religion, Wirtschaft.* München: Herbig, 1978.

Stevens, Guy, Daniel Fernando, Marc Dando & Giuseppe Notarbatolo Di Sciara. *Guide to the Manta and Devil Rays of the World.* Princeton: Princeton University Press, 2018.

Svenskt kvinnobiografiskt lexicon. *Database.* Gothenburg: University of Gothenburg, 2018 <www.skbl.se>.

Svensson, Sam. 'Bark och barkskepp.' In: *Årsbok/Föreningens Sveriges sjöfartsmuseum i Stockholm*,1944; pp. 125–146.

Svensson, Sam. *Handbook of Seaman's Ropework* [Transl. by Inger Imrie. Drawings by Kai Bäling]. London: Adlard Coles Ltd, 1971.

Svensson, Sam. *Sails Through the Centuries* [illustrations by Gordon Macfie]. Macmillan: Collier-Macmillan, 1965.

Sveriges skeppslista ['The List of Swedish Ships', published since 1869]. Published by the state institution Sjöfartsverket (the Swedish Maritime Administration); based on the list of Swedish ships, which is kept by Sjörfartsregistret (the Swedish Register of Shipping) in the same civil service department. Another important contributor to this conscientious and indispensable publication is the legal authority Domstolsverket (the Swedish National Courts Administration). Sjöfartsregistret (contributor). Domstolsverket (contributor). ISSN 0347-7258. Published and printed in Stockholm: Fritzes offentliga publikationer, 1869–1998 (series).

Tatem, Andrew J., D.J. Rogers & S.I. Hay. 'Global transport networks and infectious disease spread.' In: *Advances in Parasitology* (ISSN 0065-308X. London: Academic Press), vol. 62, 2006, pp. 293–343.

Taylor, Andrew. *Two Years Below the Horn: Operation Tabarin, Field Science, and Antarctic.* Winnipeg: University of Manitoba, 2017.

The Times Comprehensive Atlas of the World. 15th ed., London: Times Books, 2018. Maps © Collins Bartholomew Ltd 2018.

Thorman, Åke. *För passadernas vindar på 1890-talet: en skildring av livet ombord i en svensk djupvattenseglare på färd över nära och fjärran hav.* Stockholm: Nautiska förlaget, 1953.

Västerviks kyrkoarkiv, Död- och begravningsböcker. (Archives MSS.). The Book of Deaths and Funerals of the Westervik Congregation. Västerviks kyrkoarkiv, Västervik, Sweden, SE/VALA/00443/F/4 (1900-1913). 'Västerviks kyrkoarkiv (1647-1999)'. Landsarkivet i Vadstena (depå: Vadstena Castle) <https://sok.riksarkivet.se/arkiv/i93vyavqQWZTuu1eMvyUeB>.

Verne, Jules. *Vingt mille lieues sous les mers: Illustré de 111 dessins par de Neuville*. Paris: J. Hetzel, 1871. (Swedish translation: *En verldsomsegling under hafvet*, Stockholm: Pettersson, 1872.)

Wallenberg, Jacob. *My Son on the Galley*. Edited and translated by Peter Graves. London: Norvik Press, 1994.

Weibust, Knut. *Deep Sea Sailors: A Study in Maritime Ethnology*. Diss., Stockholm, Nordiska museet, 1976. [An exceptional doctoral thesis about the social life of sailing ship crews, with a special focus on maritime transport in the latter part of the nineteenth century and the desires, values, roles, strains and stresses of sailors in the merchant navy. Of particular interest here is the hostility displayed towards women on the sea, and the fact that some seamen refused to sign on or signed off if there were women on board. The tome includes an analytical summary and a useful bibliography.]

Werngren, Nils & Carl Axel Östberg. *De första världsomseglingarna under svensk flagg: kapten Nils Werngren berättar*. Karlskrona: Abrahamson, 1989. [A documentary account of the first Swedish circumnavigations.]

Widholm, Dag (ed.). *Stone Ships: The Sea and the Heavenly Journey*. Institutionen för humaniora och samhällskunskap, Högskolan i Kalmar; 2007. *Series*: Kalmar studies in archaeology, 1653-431X; 3. Published by Kalmar University, Sweden.

Willers, Uno. *Svensk sjöhistorisk litteratur 1800–1943: bibliografi*. Stockholm: Sjöhistoriska samfundet, 1956. [A bibliography of Swedish naval history edited by a former National Librarian.]

Wilson, Charles. *South Atlantic Ocean. Engraved coastal nautical chart in two sheets, compiled by J. S. Hobbs*. London: Norie & Wilson, 1884. [Besides the navigator of the Atlantic, this detailed but unsatisfactory map was used by a number of other Swedish skippers, including captain Olof Peter Reinhold Olsson (1853-1921), Råå fishing village, Raus (Scania).]

List of Illustrations

Fig. 1. *Amalia Mathilda Winroth, née Löfström. Portrait photograph 1866*. Photograph owned by the author. Private possession. (The Hallengren private Archives, Mariannelund, Sweden).

Fig. 2. *Emmy and Maria as toddlers, rescued from the city fire of 1869*. Courtesy of the Maria L. Hallengren private Collection (Hållnäs, Sweden).

Fig. 3. *Amalia Mathilda lying in state in a ship coffin in 1873*. Courtesy of the Maria L. Hallengren Collection.

Fig. 4. *Karolina Själander (head mistress) and Klara Johansson (home room teacher)*. Courtesy of Gävle Länsarkiv. A unique photograph: Headmaster Karolina Själander (standing) with form mistress and scientist Klara Johansson (sitting), who forever left their marks on the Söderström daughters. The Wilhelm Lindeberg Portrait Gallery. Photographer: Carl Berggren (1847–1897). Länsmuseet Gävleborg, public domain. XLM.U06867.

Fig. 5. *The Higher Girls' School, the magnificent creation of Karolina Själander*. Courtesy of the Länsmuseet Gävleborg. The Gustaf Swedlund Collection. Creative Commons Share Alike CC BY-NC-SA. XLM.U09787.

Fig. 6. *Emy (upper left) and Mia (second right) with fellow-students and kinsfolk in 1884 before departure. The boy is their cousin Knut Zellinger*. Photo Elise Ahlgren Gefle. Private possession. The Hallengren Archives.

Fig. 7. *Maria Söderström's travel diary on a table*. Author's photo. Private possession. © Anders Hallengren. Hitherto unpublished.

Fig. 8. *The journal begins. The first page of the travel diary*. Author's photo. © Anders Hallengren. Hitherto unpublished.

Fig. 9. *'THE Atlantic. Commanded by Captain J. A. Söderström. Oil by H. Petersen & P.G. Holm, 1877'. The barque with all square sails and for-and-aft rigged mizzen sails set. The efficiency of the fast-sailing merchantman in various weather conditions and winds could be improved by extra gaff sails, trysails, or staysails—as can be seen from the oil painting. She is carrying the Swedish–Norwegian Union colours along with flying the HNJB signal flags, the ship's identification.* Private possession. The Hallengren Archives. Original courtesy of the Yngve Herrström estate and the Stig Herrström private collection. Original photograph in the author's private archives.

Fig. 10. *The enrolment register of the Atlantic in 1885.* Copy from a ledger in the Gefle mercantile marine office in the Swedish National Archives; public domain. Copy in private possession.

Fig. 11. *The 1885 crew of the Atlantic pictured in Melbourne in March 1886.* Courtesy of the National Maritime Museum, Stockholm. Fo24248A. The crew of the *Atlantic* photographed upon their arrival to Melbourne in March 1886. The personnel living in the stern castle (upper right) have had opportunity to compose their features and dress up to the occasion, whereas the rest of the staff is still in their regular working-clothes and worn after the long trip. *Standing, from the left*: Carpenter Per Johan Blomgren (b. 1844, 32 years, married), ordinary seaman C. J. Holm (b. 1865, 21 years), deckhand Knut Axel Boman (b. 1869, 17 years), ordinary seaman F. O. A. Gustafsson (b. 1865, 21 years), ordinary seaman Per August Hörberg (b. 1865, 21 years), deckhand Gustaf Theodor Lysander (b. 1868, 18 years), ordinary seaman A. E. Hanner (b. 1866, 20 years), able seaman Anders Öman (b. 1861, 24 years), deckhand Johan Christian Blomgren (b. 1866, 20 years), the steward and chef Carl Östling (b. 1848, 37 years, married), second mate Johan Leonard Clase (b. 1850, 35 years), Emmy Söderström (b. 1868, 17 years), Maria Söderström (b. 1867, 18 years), first mate Pehr Gustaf Elfström (b. 1854, 31 years, married), and captain J. A. Söderström (b. 1841, 45 years). *Sitting, from the left*: Able seaman and sailmaker Johan Gustaf Bergström (b. 1838, 48 years), able seaman Petrus Södergren (b. 1862, 23 years), ordinary seaman F. Bredenberg (b. 1868, 18 years), the cook Johan Axel Waltner (b. 1864, 22 years), and the sailmaker and seaman Johan Gustaf Sahlin (b. 1846, 40 years, married). Finally, the third female hand, the helpful ship's hound Harriet, a St. John's water dog, an extinct breed. The monthly salary of the employees varied widely, from 15 kr for deckhand Lysander to 70 kr for chief officer Elfström, but food and lodging were included with the round-the-clock attendance.

Fig. 12. *Maria Söderström, the diarist, in Australia in 1886.* Courtesy of the Maria L. Hallengren Collection. Original photograph of 1886 carefully restored and coloured by Maria L. Hallengren. Also used on the cover.

Fig. 13. *Emmy Söderström on the voyage.* Courtesy of the Maria L. Hallengren Collection. Original photograph of 1886 carefully restored and coloured by Maria L. Hallengren. Also used on the cover.

Fig. 14. *Captain Cutler and his daughter.* Courtesy of Colin O'Neill. Private possession.

Fig. 15. *The Dufour map of 1863. Passing south of Australia outward and by northern roundabout homewards was a challenge and called for many stops. The Atlantic tried these routes several times, choosing various alternatives.* Private possession. Cartographic material owned by the author.

List of Illustrations   299

Fig. 16. *Captain John Ljungberg, Maria's great love.* Private photograph. Courtesy of the Maria L. Hallengren Collection.

Fig. 17. *To the crew of the Atlantic, the important destination Malden Island was in the middle of nowhere. But they were not alone. The Dufour map of 1863 used by captain Söderström. Detail.* Cartographic material owned by the author.

Fig. 18. *The Skipper King. Captain Johan Axel Söderström in his prime.* Photograph in private possession. Original including negative in The Hallengren Archives.

Fig. 19. *Captain Axel Söderström portrayed by Christophe Delphin Comergnac in Rochefort harbour, France, in 1882.* Private possession. Original including negative in The Hallengren Archives.

Fig. 20. *Nautical Chart, the Azores (detail). Perilous waters feared by the captain.* SK 025 Swedish National Maritime Museum. Copy in private possession.

Fig. 21. *a-d = H N J B. The HNJB signal flags, the Atlantic's identification.* Copies in private possession. Also in: Wikimedia Commons, the free media repository / Creative Commons Share Alike.

Fig. 22. *The Princess' Theatre, Melbourne.* Photo Anders Hallengren 27 January 2009. Private possession.

Fig. 23. *Royal Exhibition Building, Melbourne, built in 1880.* Photo Anders Hallengren 28 February 2009. Private possession.

Fig. 24. *The Atlantic resting in harbour. The barque Atlantic was built at O. A. Brodin's Shipyard in Gefle in 1876.* Courtesy of the Länsmuseet Gävleborg Photo Collection. Identifier XLM.A98-34.

Fig. 25. *Maria Söderström as a bride on 28 July 1888.* Courtesy of the Maria L. Hallengren Collection.

Fig. 26. *John Ljungberg's death certificate. Cape Town 22 March 1901.* Private possession. Copy in the author's private archives. Official transcript owned by dir. Thorleif Herrström, Stockholm. Source: A F Clarkson, Western Cape Archives, Republic of South Africa.

Fig. 27. *The family memorial in Westervik. Maria Söderström raised the stone to the memory of beloved father, husband, and daughter; but her spouse was interred in South Africa. When her only son died in 1929, she added his name to the black stone of diabase, now a relic of culture at the Old Cemetery.* Photo: Maria L. Hallengren. Courtesy of the photographer. The Maria L. Hallengren Collection.

Fig. 28. *The Atlantic in full sail, 'Captain O. G. Nilsson,' 1910.* Public domain. Courtesy of the National Maritime Museum, Stockholm. Fo80338A.

Fig. 29. *The Gavle River harbour with the Atlantic and other ships in 1899.* Courtesy of the National Maritime Museum. Fo24989AB.

Fig. 30. *The last logbook of the Atlantic. Cover Page.* Kommerskollegium, Huvudarkivet, series F III d 'Skepps- och Maskindagböcker (1891–1929)', vol. 9. (Riksarkivet, The Swedish National Archives, https://riksarkivet.se).

Fig. 31. *The last logbook of the Atlantic. Cover.* Private copy.

Fig. 32. *The Mo Harbour, a familiar destination to the Atlantic, now no longer found on nautical charts.* Copy in the Hallengren Archives.

Fig. 33. *Mia, the writer and traveller, in her old age among heirlooms, pictures, and memories.* Photograph in private possession. Courtesy of the Maria L. Hallengren Collection.

Fig. 34. *Emmy's daughter Greta Videnstam took charge of the leading shipping office and became shipbroker.* Photo Carl Larsson 1932. Portrait photograph in private possession. The Hallengren Archives.

Fig. 35. *Emmy with her family of nine in the Central Palace, Gefle, c. 1910.* Courtesy of the Maria L. Hallengren Collection. From the left: Sven, Gunborg, Emmy, Greta (the firstborn), Bo, Ingrid (the youngest), Gotthard (husband), Tord, Stig.

Fig. 36. *Emmy's home became a centre for musical life. Her husband with his Gefle String Quartet in 1932.* Photo original owned by the author. Private possession.

Fig. 37. *The Hallengren & Herrström catalogue of 1897.* Copy in private possession.

Fig. 38. *The polar explorer Otto Nordenskjöld passed his only student in geography, Emmy's son.* Extant examination book in the author's possession. Authors' photo.

Index

A

Aborigines of Australia, 207
Adams, John, 86, 87
Adelaide, 55, 56n18, 61–69, 66n28, 67n31, 68, 72, 73, 119, 137, 139, 190, 191, 259
Adelaide Botanic Garden, 66
Africa, 165, 225, 232
Afrika, Russian warship, 78n60
A Fugitive Crosses his Tracks, 169
Ahlgren, Johan, 121n92, 122
Aitutaki, 148, 149, 206
Åkesson, Johan, 254
Albatross, The, 6, 59–61, 84n65, 225, 226
Albert Dock, 252
Alcides, full-rigged ship, 177
Alcúdia, 203
Alighieri, Dante, 126, 159
Almería, 235
America (USA), 14, 74, 75, 77, 80, 96, 99n77, 100n79, 103, 113n87, 117, 129, 141, 166, 170, 200, 205, 214, 221, 235, 239, 241, 257, 272
American Indians (see: Native Americans)
Andalucía, 235
Andersson, Eric Engelbert, 11
Andrea, full-rigged ship, 19, 185
Ankarcrona, Sten, 201
Anna Karenina, 201, 202
Annerstedt, Carl, 189
Antarctic, barque, 240, 280
Antarctic, the, 8, 57, 91n68, 119, 167, 223, 280, 281
ANZAC, 258, 259
Aotearoa (New Zealand), 207
Aquidneck, barque, 200, 229–230
Argentina, 214, 220–222, 228
Arvid Nordquist, 277
Åsbrink, Carl Olof, 17
Asiatic Cholera epidemic, 1, 11, 26, 41n2, 200, 267
Atatürk, Kemal, 258
Atlantic, barque, xii, 4, 7, 8, 37, 40, 41n2, 43, 44n6, 47n10, 52, 54n16, 57, 61–63, 65, 68n34, 69, 70, 81, 88, 94n72, 97, 99n78, 100–102, 104, 105, 111, 116, 117, 120, 123, 131, 138, 142, 144, 153, 154, 156n109, 161, 166–168, 171, 180, 181, 183–191, 199–201, 204, 206, 211, 213–215, 221, 222, 224, 227, 228, 230–233,

235, 240–245, 249, 250,
  252–254, 267
Atlantic Ocean, the, 2, 21n3,
  44–45, 49–53, 57, 143, 159,
  163, 165, 166, 184n4, 194,
  240, 241, 258, 263
Atlantic's ship's boat, 151, 249
Audacia, full-rigged ship, 15
Aurora, barque, 69
Australia, xiii, 5, 54, 55, 57, 61–69,
  68n33, 69, 70, 71n45, 74, 76,
  78n60, 82n100, 90, 97, 99n78,
  103, 105, 116–118, 121, 124,
  137, 138n97, 141, 143–145,
  149, 151, 152, 166–168, 170,
  190–192, 200, 205, 207, 214,
  222–226, 236, 239, 256–261,
  265, 272
Azores, 24, 163, 186, 230,
  240, 241

B

Baía de Santos, Brazil, 235
Baltic, brig, 190
Baltic Sea, the, xiin1, 5, 20, 43, 62,
  79n61, 140, 168, 241, 243,
  250, 251
Banda Sea, 142
Bång, Lars, 12, 17
Baranov, A., 78, 79, 79n61
Barbados, 235
Baret, Jeanne, xiii
Bass Strait, 120, 121
Baudelaire, Charles, 60
Bavaria, 13
Beagle Channel, 159
Beckton Gas Works, 250
Beijbom, Ulf, 257
Berge, Lars, 177, 178

Bergendal, Gunborg Emmy Maria,
  5, 270, 274
Bergström, Johan Gustaf, 63, 194
Birkenhead dock, 218
Bizerte, 235
Blackwell, 249
Bligh, William, 85, 149, 167
Blixten, frigate, 184
Blomberg, J. Gustaf, 190
Blomgren, Johan Christian, 63
Blomgren, Per Johan, 63, 91,
  204
Boccaccio, 71, 72, 207
Boccherini, Luigi, 270
Boman, Knut Axel, 63, 194
Bonne Amie Flygare, 34
Bonne Amie Sundström, 34
Book of Jonah, 177
Bordeaux, 144, 190, 277
Borg, Elsa, 34
Brazil, 235
Bredenberg, F., 63
Bristol, 165, 190
British Board of Trade, 220n10
British Columbia, 99n77, 202
Brodin, O. A., 21, 37, 143, 187,
  188, 211, 241, 243
Brown, Alura Eliza, 106, 107
Brynäs Wharf, 228
Buenos Aires, 57n20, 214, 217,
  220, 221, 228
Bulgaria, 200

C

Callao, full-rigged ship, 191
Callao, seaport, 170, 195
Camelot, 6
Canada, 99n77, 112n87, 202, 203,
  216n5

Canadian Pacific Railway, 202
Cape Horn *(Cabo de Hornos)*, 153–159, 167, 170, 214, 221, 223–227, 280
Cape of Good Hope *(Kaap die Goeie Hoop)*, 60, 166, 168, 204, 205
Cape Town, xiii, 8, 79n61, 231–235, 237
Cape Verde Islands (Cabo Verde), 48, 166
Captain Ahab, 172
Captain Nemo, 169
captain's mansion, 19n1
Cardiff, 21, 190, 195
Caribbean, 26, 165
Carina, 90
Carle Herrström, Pontus, 277, 279
Carlsson, Frans Arnljot, 267
Carlton, Victoria, 140
Carter, Cara, 110, 114
Carthage, 163
Casparsson, Edvard, 31
Cassiopeia, 90
Castletown, 161
Cedermark, Alvar, 274
Centaurus, 90
Central Palace, Gefle, 272
Cepheus, 90
Chapeau de Maître, 121
Charles's Wain, 89
Chicago, 235, 262
Chief Chetzemoka, 100n79
Chile, 143, 145, 153, 170, 221, 223, 280
Chili, barque, 142, 143, 150, 151
Chimakum, 74n53, 100n79
China, 74n52, 103, 163, 168, 170, 188, 200, 207, 222
Chincha Islands, 170n2
Chinook, 100n79

Chittagong, 163
Christchurch, NZ, 206
Christina, barque, 19, 184, 185
Clallam, 74n53
Clapp, G. P., 113
Clari, or the Maid of Milan, 129
Clase, Anders Leonard, 18, 21, 26, 47n10, 181n3, 185
Clase, Elma Charlotta Leontina Andersdotter, 68n35
Clase, Johan Leonard Andersson, 63, 68n34, 73n50, 192, 215n4, 216, 217n6
Clase, Johanna Katarina, née Söderström, 18, 26, 47n10, 68n34, 192
Coke Company in Beckton, 250
Coleridge, Samuel Taylor, 59
College of Navigation, 185
Columbus, Christopher *(Cristoforo Colombo)*, 165, 166, 194, 262, 263
Conflagration of the City of Gefle in 1869, The, 22–24, 227, 273
Conrad, Joseph, 196n8, 199
Cookery for Seamen, 219n8
Cook Islands, 149
Cook, James, 149, 159, 166, 188
Cook's Bay, 194
Cooper ridge, 92n69
Copp, Dolly, 112
Copp, William Harvey, 112n87
Coriolis, 57
Coriolis, Gaspard Gustave, 173
Cove's harbour, 144
crown (krona), Swedish coinage, 31n6
Crux, 90
Cutler family, 108, 109, 112

Cutler, Rosewell Dwight, 106, 107, 112
Cutler, Ruth, 106, 107
cutter, 25
Cutty Sark, 168

D

Daniel, barque, 185, 267
Daniel Elfstrand Pehrsson, full-rigged ship, 170
Daphne, barque, Captain Johan Pettersson, 144, 150
Dardanelles, 258
Delagoa Bay, 160
de Luze, Gwénola, 279
Dickson-Waern, James, 82n64
Diodorus Siculus, 164
Dockenhuden, barque, 138n97
Dogger Bank, 144
Doldrums, the, 91, 92n69, 183, 265
Doyle, Sir Arthur Conan, 24, 200
Draco, 90
Drake, Sir Francis, 159
Dr No by Ian Fleming, 150n103
Duwamish, 100n79

E

Earl Granville, barque, 112n87, 113
East Coast Fishing Disaster, 144
Easter Island, 148, 152n104, 194
Eggegrund lighthouse island, south Bothnian coast, 247
Egypt, 163, 164, 258
Elfström, Johanna Christina, 195

Elfström, Pehr Gustaf, 42, 53, 63, 77n58, 94, 104, 192, 194–196, 204, 213
Elsinore (Helsingør), 17, 43–44, 61, 69, 250, 253
Emerson, Ralph Waldo, 202, 275
Emperor Alexander III, 79n61
Endemic tropical fever, 262
England, 65n27, 71n45, 85, 87, 160, 161, 202, 223, 229, 250, 251
English Channel, the, 44, 158, 161, 165, 166
Engström & Luth, 186, 188
Engström, Per Adolf, 185, 188, 215, 231, 233
Ericsson, Eric, 184n4
Erik the Red, 194
Estates of the Realm, 239
Euclid, 32, 179

F

Fårö, 251
Fiji, 141, 145
First Fleet, 209
Florida, 228, 235
Flying Dutchman, 205
Ford, Bessie, 65
Forslund, Johan Erik, 13
Fourth Quarter, 13
France, 184, 200, 214, 225, 258
Francotte à Liège M/1871, 181
Freycinet, Rose de, xiii
Freygeirr, Viking Chieftain, 169
frigate birds, 6, 84n65, 145
Fryxell, Cecilia, 34
Fuller Ossoli, Margaret, 202
fulmars, 6

G

Gädda, Karl Henrik, 69n38, 73, 74
Garberg, Rickard Wilhelm, 25, 26, 143, 171n2, 213–219, 220n9, 226, 228, 229
Garberg, Sophie, 34
Garberg, Sven, 215n4, 220n9
Gästrikland, 10
Gauguin, Paul, 64n25
Gavle River, 4, 242
Gefle City's First District, 17
Gefle City's Second District, 11, 30
Gefle Gasworks, 229
Gefle mercantile marine office, 52, 185, 243
Gefle (Gävle), Swedish seaport by the Baltic Sea, xiin1, 10, 11, 13, 14, 20, 21, 29, 32, 34, 41, 52, 61, 66, 79, 150, 166, 168–171, 185, 187, 188, 190–192, 196n8, 198, 199, 211, 214, 227, 228, 230, 233, 235, 241, 243, 247, 250, 267, 268, 274, 279
Gelibolu (Gallipoli), 258
German Club, 78
Germany, 54n17, 125, 167, 200
Gevalia, barque, 137
G. H. Mumm & Cie, 279
Ghosts (*Gengangere*), 262, 264
Gig Harbor, 214, 221
Gilbert & Sullivan, 81n63
God, 3, 4, 34, 41, 42, 45–50, 53, 55, 58–60, 65, 67, 71, 73, 79, 80, 83–85, 87, 88, 90, 95–97, 100, 102, 108, 115, 117, 118, 123, 136, 137, 140, 141, 143, 146, 147, 154, 155, 157, 158, 160, 161, 178

Gorshkov, A., 79n61
Gothenburg, barquentine steamship, 65
Gothenburg, seaport on the Kattegat strait, 5, 70, 139, 161, 169, 271, 281
Gotland, 41, 203, 251
Granström, J.W., sailor, 177
Grape, Sophia, 18
Gravesend, 249
Great Barrier Reef, 65
Great Tea Race, 188
Greenland, 164, 194, 216n5
Grubbström, Isak Petter, 54, 70, 72, 118
Grundén, Julia, 119
Grundén, Mary, 77, 81, 82, 91, 93–95, 98, 102, 105, 106, 115, 116, 118–120
Guangzhou, 163
Gulf of Bothnia, xiin1, 41–44, 61, 168, 182, 248
Gulf of Mexico, 235
Gulfport, Mississippi, 240
Gunnersen, C., agent, 61, 99n78
Gurli, barque, 70n44, 139–141, 161, 208, 211, 212, 213, 229, 232, 261
Gustafsson, F. O. A., 63
Gustaf Wasa, full-rigged ship, 19, 21, 24, 181n3, 185, 243

H

Hägg, Aquilina, 267
Hägg, Jacob Adolf, 267
Hägg, Jacob, rear-admiral, 267
Hallengren, Bo Hilding, 227, 268, 270, 271, 274

Hallengren, Gotthard, 227, 237, 255, 267, 273, 277, 279
Hallengren, Jacob, councillor, 1, 227, 267
Hallengren & Herrström, 277–280
Hallengren, Johan Jacob, 277–280
Hallengren, Stig, 227, 272, 274
Hallengren, Sven Gotthard, 239, 255, 266, 271, 274, 281, 282
Hamburg, 152, 160, 161, 190, 211
Handeland, Lars, 261
Hanner, A. E., 63
Hanseatic League, 201, 231
Härjedalen, 10
Härnösand, 232, 233, 250, 256
Harriet, ship's dog, 63, 94n72, 103, 215n5
Hartlepool, 250
Hawaii (Sandwich Islands), 69, 107, 109, 111, 152n104
Hazard, schooner, 15, 17
Hecataeus, 164
Heidsiek & Co, 279
Henricson, Ingvar, 257
Herodotus, 164
Herrström family, 277
Herrström Island in the Weddel Sea, 280
Herrström, Oscar (1857–1923), 277, 279, 280
Herrström, Stig Torolf Einar, 261, 279
Herrström, Thorleif, 233
Herrström, Torsten Bure, 182
Herrström, Yngve Axel Einar, xv, 70n44, 261
Higher Girls' School, 34, 35
HM *Bark Endeavour*, 188
HM Corvette *Freja*, 187
HMS Blonde, 145

HMS Bounty, 85–87, 88n67, 149, 167
HNJB signal, 40, 144, 189
Hobson's Bay, 99n78, 117, 120
Hoh, 74n53
Holm, C.J., 63
Holy Grail, 6
Holyhead lighthouse, 218
Honolulu, 69, 107, 110
Hopp, pewterer, 7
Hoppet, frigate, Anders Osenius and Henric Martin Captains, 184n4
Hoppet (The Hope), schooner, 184
Hörberg, Per August, ordinary seaman, 63
Hotel Söderström, 214, 220
House of Nobility Street, 19
Hudiksvall, 230, 231
Hummel, Alma, 55, 56, 65, 66, 72, 119, 122–124
Hummel family, 56n18, 65
Husqvarna armourer's workshop, 181

I

Ibsen, Henrik, 201, 262
Iceland, 164, 194, 216n5
Idogheten ('industry'), brig originally named Napoleon, 13
Illis quorum, 37
India, 163, 170, 198, 199, 230, 249
Indian Ocean, the, 57, 60, 121, 168, 204, 258
Indian Shaker Church, 103n82
Indonesia, 177
Intertropical Convergence Zone, 92–96

Isla Herrstrom, 280
Island of Malden, 150
Islandsholmen, 15, 17
Isle of Portland, harbour, 160–161, 161n114

J

Jacobs, Henry, 206
Jamaica, 2, 25
James Ross Island, 280
Japan, 81, 201, 255
Java, 142, 198
Jawa, full-rigged ship, 24
Jehovah, 177, 178
Jesus Christ, 41, 42, 45, 46, 49, 51, 55, 58, 80, 84, 87, 89, 95, 96, 98, 115, 122, 123, 146, 147, 155, 157, 180
Joe Hill, 196n8
Johansson, Carl Ludvig, 213
Johansson, Klara, 33, 34, 72, 131n96
Jönköping safety match factory, 198
Josephson, Ingrid, née Hallengren, 270, 274
Juan de Fuca Strait, 74n53, 99n77, 221
Juan Fernández (Robinson Crusoe Island), 221
Juno, brig, 44
J.W. Setterwall, ship, 186

K

Kalākaua, the last king of Hawai'i, 111
Kanaguri, Shizō, 255
Karling, Maritha, 266

Kildahl, Norwegian family, 101, 104, 109, 110, 113, 114
King Carolus XII, 155n107
King Gustaf V, 256
King Island, 121
Kingston, 25
Kingston upon Hull, 235, 251, 252
Kipling, Rudyard, 230
Kiribati, 149, 206
Kiteley, William, 110, 113, 114
Kitsap County, 100n79, 103n82
Klickitat, barkentine, 107
Klintberg, Anders Gustaf, 29, 30
Klintberg, Carl, 29
Klintberg, Clara Mathilda, 29, 30, 62, 64
Korsnäs AB, 228
Krakatoa (Krakatau), 177
Kruzenshtern, four-masted barque, xiv
Kubikenborg, 231

L

Labrador, 203, 216n5
Lady of St Kilda, schooner, 120
Lancastrian method of mutual instruction, 18
Land of Punt, 164
Långbro mental hospital, 265
Lang, Waldemar, 79n61
La Royale, French man-of-war, 201
Le Havre, 195
Leif Eriksson, 194, 216n5
Leitmotif, 273
Levin, Lully, 50, 64, 72, 79, 92, 110
Liberation, 32–37
Lindberg, Johanna, 12, 18, 28
Lindblad, Adolf Fredrik, 130

Lind, Clara, 34
Lind, Jenny, 202
Lindos, 203
Lindström, Carl, 73
Lindström, Carrie, 66, 72
Lindström, Ottoline Louise, m. Ramenius, 66, 72
Line Islands, 141, 145, 166, 272
Line, The, 53, 61, 92
Linklater, Eric, 186
Linnaeus, Carolus, 56, 262
Little Boston, 100n79, 103, 108–111
Liverpool, 213–215, 217n6, 219n8, 220, 228
Lizard Point, the southernmost point of the British mainland, 61, 144, 190
Ljungberg, Axel Erik Douglas, 213, 232, 237, 238, 256–261
Ljungberg, Dolly Barbro, 231, 232, 237, 238, 256, 261
Ljungberg, Gurli Maria, 229, 232, 237, 256, 261, 264, 265, 277
Ljungberg, Johan 'John' Erik, 70, 124, 136, 138–141, 146, 147, 153, 157, 160, 208, 211, 213, 229, 232–234, 236, 237, 256
Ljungberg, Sven Reinhold, 140, 233
Ljung, Eric Andersson, alias Captain Young, 186
Ljusne, 26, 190
Lobos de Tierra, 170n2
Loch, Henry Brougham, 1st Baron Loch, governor, 74n52
Löfström, Carolina Petronella, 10, 18
Löfström Hammar, Carl Johan, 10, 14, 28

London, 25, 64n25, 75n57, 165, 168, 200, 241, 247–250
London Bridge, 25
Lord Byron, 145
Lues venerea (Syphilis) as a familial affliction, 261–263, 266–267
Luleå, 69n38, 250
Lundberg, Dan, engineer, 66
Lysander, Gustaf Theodor, 63, 194, 203

M

Madame Bovary, 201
Madsen, Artur, 253
Magellan (Magalhães), 5, 159
Magellan, barque, 195
Maitland Cemetery, 233
malaria, 262
Malden Island, xiii, 136, 141–149, 151, 154n105, 159, 206
Malmö, 252
Maori, 206, 207
Marconi, Guglielmo, 2
Maria, brig, 34
Martin, John Stanley, 257
Mathilda, barque, 21
Mauritius, 199
McPhee, Eunice, 109, 113
Mediterranean, 21, 163
Megasthenes, 164
Melanesia/Melanesian 152n104, 206
Melanesia, barque, 160
Melbourne, 7, 62, 63, 64n25, 66, 67n31, 69–76, 71n45, 75, 77, 78n59, 80, 81, 92, 97, 99n78, 103, 109, 116–121, 136,

Index **309**

138n97, 139–142, 145, 148, 150, 151, 160, 194, 200, 205, 207–209, 211, 214, 223, 257
Melville, Herman, 60
Mercedes automobiles, 239
Mexican Gulf, 228
Micronesia/Micronesian, 136–143, 149, 206
Mikado, 81, 200, 207
Milky Way, 90, 175
milréis, 240, 241
Miltopaeus, Andrietta, 21n3
Miltopaeus, Fredrik W., 21n3, 181n3
Miltopaeus, Hjalmar, 21n3
Minerva, barque, 15, 19n2, 184
Mineur, Claes Johan, 196n8
Mitzlaff, Eugenia von, 51n13
Moberg, Göran, 243
Moberg, Karl Johan, 30
Moby-Dick, 60
Modig, Norwegian steamer, 261
Mo Harbour, 248, 249
Monarch, barque, 191
Montevideo, 57
Moody, Dwight L., 65n27
Moray Firth Fishing Disaster, 144
M/S Colombia, 182
Mud Islands, 78n60
Murray, Andrew, 42n4
Mutiny on the Bounty, xiii, 84–89

N

Nanna, barque, 54, 70, 72, 118
Nathaniel Johnston & Fils, 279
Native Americans, xiii, 74n53, 99, 100, 100n79, 102, 103n82, 107–108, 194, 206

Navigation School, 19, 241
Nephritis, 233, 237
Neptune, 53, 54, 93, 224
New Brunswick, 112
New Caledonia, 141, 145, 152n104, 206, 272
Newfoundland, 216n5
New Guinea, 90, 142, 272
New South Wales, 259
New York, 21n3, 25, 75n57, 202, 271
New Zealand, 70n41, 80–82, 90, 97, 98n75, 141, 152n104, 166, 170, 206, 207, 258, 259, 272
Nilsson, Christina, 202
Nilsson, Oskar Gustav, 241–243, 245, 247, 251, 253
1912 Summer Olympics, 255
Niue, 148, 149, 206
Non multa sed multum, 32
Non scholæ sed vitæ discimus, 37
Nordbladh, Carl, 29, 30
Nordenskjöld Coast, 281
Nordenskjöld, Otto, 240, 280–282
Norfolk Island, 88
Norrköping, 260
Norrland, 10
North America, 99n77, 103n82, 164, 166, 184
North Sea, 44–47, 61, 67, 144, 166, 171, 185, 204, 241, 243, 252
Nyström, Johanna Margareta, 14

O

Öberg & Horndal, shippers, 254
Öberg, Pehr Johan, 17
Oberlin, brig, 74n51
Ocean, barque, 15

Oceania, xiii, 166, 168, 272
octant, 17
Oden, barque (built 1849), 15
Oceanic bonito (Skipjack tuna), 56
Oest, Eva Henrika, 279
Öland, 42
Olofsdotter, Anna Greta, 10
Olsson, Otto, 235, 240
Olympic Peninsula, 103n82
Öman, Anders, 63
Onassis, Aristoteles, 268
Øresund, 41–44, 250
Orient, barque launched in 1884, 121
Örnsköldsvik, 241
Orphanage of Salem, the, 4, 14–18
Oscar II, king of Sweden and Norway 1872–1905, 68n33, 171
Osenius, Anders, 15, 29
Österlövsta, Upland, 10
Öster—'the East,' 10
Östling, Carl, 48, 49, 63, 92, 93, 194, 204
Otago, barque, 199
Ovanåker, 68n36

P

Pacific Ocean, the, xiii, 6, 77, 77n58, 84, 103n82, 112n87, 116, 117, 144, 148, 152n104, 159, 163, 167, 180, 205, 206, 221
Pakistan, 163
Pamir, barque, 167
Papua, 90, 206
Patagonian ice fields, 281
Patton, George S., 255
Patu-iki (king) Tui-Toga, 149

Peru, 143, 170, 195
petrels, 6, 84n65, 226
Pettersson, Hanna, 47, 50, 64, 88, 104
Pfeiffer, Ida Laura, xiii
Phoenicians, 163, 164
Phthisis plague (tuberculosis), 2, 26, 158n111, 267
Pirates in the Deep Green Sea, 186
Pirates of the Caribbean, 44n6
Pitcairn Island, xiii, 84–89
Piteå, 251
PJ Hægerstrand company, 121, 227–229, 231, 267, 268
Plato, 163, 258
Platon, Russian man-of-war, 78n60
Pleiades, 207
Plough, 90
Point Lonsdale, 78n60
Point Nepean, 78n60
Polar Star, 173
Polynesia/Polynesians, xiii, 148, 149, 152n104, 163, 180, 206
Pompey the Great, 173
Pomponius Mela, 164
Popov, A, 79n61
Port Campbell, 205
Port Gamble, 99, 100n79, 101–104, 107, 111, 112, 114
Port Germein, 214, 223
Portland cement, 161
Port Ludlow, 109, 110
Port of Callao, 170n2
Port of Drammen, Norway, 70, 192
Port of Gävle (Gefle), 41
Port of Pensacola, 228, 235
Port Phillip, 75n55, 78n60, 120
Port Pirie, 141
Portsea, 78n60

Port Townsend, 74, 96–98, 98n75, 99–101, 105, 107, 109, 111–117, 214, 221
Portugal, 240, 263
Prince Gustav Channel, 280
Prince of Wales Opera House, 72, 75n57
Prince Oscar Carl August Bernadotte, Count of Wisborg, 110, 111
Princess' Theatre, 71n48, 81n63, 207, 208
Puget Sound, 74, 98n75, 99n77, 100n79, 214, 221
Pythagoras, 179

Q

Queen Lili'uokalani, 111
Queenscliff, 78n60

R

Rabén, Karl, 279
Rabén, Sully, 279
Ragnar, schooner, 169
Rapa Nui, 148
Rarotonga, 149
Rederi AB Svea, 261
Regent's Canal, 250
Rettig P(er) C(hristian) & Co, 198
Rettig, Robert, 1818–1886, 21n3
Rhodes, 203
Riksdaler (rd.), 31n6
Rip Van Winkle, 75, 207
River Mersey, 213
River Plate, 57n20, 214n3, 220
Robinson Crusoe, 221
Rochefort harbour, France, 184

Römcke, Otto, 70, 71, 77, 78, 99n78, 120, 138
Rothschild, David C.H., 100
Rothschild, Dorette, 100
Rothschild, Ernest Eugene, 100
Rothschild family, 99–100, 103n82, 111, 114
Rothschild, Louis, 100, 101, 109, 111, 116
Rothschild, Regina, 99–101, 105, 108, 109, 112, 115, 116
Rouen, 214, 225
Royal Botanic Gardens, 73
Royal Exhibition Building, 71n48, 73, 141, 207, 209
RSL (Returned and Services League of Australia), 259
Rückert, Friedrich, 131
Rumford, Sir Benjamin Thompson, Count, 13
Rumford soup, 13
Russia, xiii, xiv, 170, 200, 201, 255, 258, 273
Russian Armada, the, 5, 77–80

S

Sabine City, 230
Sabine Pass, Texas, 230n13
Säfström, Levin, 24
Sahlin, Johan Gustaf, 63
Saint Helena, 18, 24, 47n10, 68n34, 97n74
Sal anglicum, 220
Samoa, 90, 145, 149
Sandell, Lina, 51n12, 58n21, 115n90, 122n94, 157n111
Sandemose, Aksel, 169
Sanders family, 111

Sandridge Pier, 67n31, 69, 80, 97, 205
Sandström, Mathilda 'Thilda,' née Löfström, 28
San Francisco, 75, 107, 109, 112, 221, 230
Santo Antão, 48
São Miguel, 241
Sätterlund, Birger Gotthard, 69, 74, 121, 124
Schierbeck, 43
Schubert, Franz, 273
Seamen's Church, 24
Seattle, 74n53, 103n82, 107
Sea-Wolf, 172
Seferis, Giorgos (Georgios Seferiades), 179
Selkirk, Alexander, 221
Senegal, 154n105, 191
Señora Santa María del Buen Ayre, 221
Serbia, 200
sextant, 2, 17, 69, 178
Shakespeare, William, 188
Shanghai, 168
shearwaters, 6, 84n65
ship as symbol, 27
Ship's cat, 94, 94n72
Shipwreck Coast, the, 120
Sindbad, 174
Singapore, 199
Sinuhe, 174
Sirén, schooner, 184
Själander, Karolina, 32–35, 37, 59, 192
Själander, Per Martin, 34
Sjöström, Gustaf Adolf, 137, 138
Skagen Lighthouse and the Skagen Painters, 253
Skanør–Falsterbo, 251

Skipper King, the, 3, 171, 172, 236–239, 265
S'Klallam Tribe, 100n79
Skokomish, 100n79, 103n82
Skönvik, Sundsvall, 241
Skutskär, 10, 41, 66n30, 192
Slocum, Joshua, 177, 200, 229–230
Södergren, Petrus, able seaman, 63
Söderhamn, 191, 230, 231
Söderström, Carl Gustaf, 18, 44, 144, 185
Söderström, Carolina Josefina, 18
Söderström, Emmy 'Emy' Axelina Charlotta, xiii, 2, 4, 10, 21–24, 27, 28, 30, 32, 36, 37, 48–50, 52, 54, 57, 59, 63, 64, 72, 76, 79, 92–94, 96, 105, 106, 115, 124, 151, 153, 154, 161, 171, 172, 183, 192, 193, 195, 200, 201, 203–209, 213, 214, 216, 227, 229, 231, 237, 239, 255, 257, 261, 265, 267, 268, 270, 272, 274, 281
Söderström, Erik, 22
Söderström, Johan Axel, xii, 2–4, 7, 18–21, 24–26, 29, 37, 40, 42, 45, 47n10, 50, 51, 55, 56n18, 57, 61–64, 66, 68n36, 69, 71, 72, 74, 78, 79, 88, 92, 93, 95, 105, 109, 111, 112, 115, 118, 123, 124, 139–141, 144, 153, 154, 156, 158, 160, 161, 166, 170n1, 171, 172, 175, 181n3, 183–186, 188, 190–192, 194, 207, 213, 214, 220, 225–232, 236–238, 240, 241, 243, 256, 261, 267, 272
Söderström, Jonas Reinhold, 19n2, 144, 184, 237

Söderström, Maria 'Mia' Mathilda
 Charlotta, xiii, 4, 5, 7, 20–23,
 27, 28, 30, 32, 34, 36, 37, 39,
 48, 63, 64n25, 68n36, 70n44,
 76, 79n61, 129, 166, 191, 192,
 200–202, 207, 211–214, 229,
 231, 232, 236–239, 256, 260,
 261, 264–267, 273, 275, 277,
 279, 281
Solid, barque, 191
Somerset Hospital, 233
Sophie, full-rigged ship, 25
Sophie, schooner, 15, 17
South America, 57, 144, 145, 152,
 195, 221, 225, 280
Southern Cross, 89–91, 173
Southern Ocean, 8, 57, 60, 61
South Island, 98n75, 206, 272
South Seas, the, 64n25, 84, 86, 145,
 149, 151, 152n104
Spain, 21, 45, 235, 263
Spencer Gulf, 214, 223
Spice Islands (Maluku), 142
Square Grand Piano, 18, 31, 32, 52,
 273
S/S Stockholm, 271
Statue of Liberty–Ellis Island
 registry, 14
Stedman, Ingrid Gurli Maria, née
 Herrström, 20, 264
Stella Polaris, 90
Stevenson, Robert Louis, 44n6
St. George's Channel, 218
St Kilda, 120
Stockholm, 7, 11, 14, 34, 71, 164,
 172, 186, 197, 199, 200, 202,
 241, 249, 250, 255, 260, 264,
 265, 277, 280
Stora Kopparbergs Bergslags AB,
 190

Stora Teatern, Gothenburg, 271
Stowe, Harriet Beecher, 202
St. Petersburg, 79n61
Strabo, 164
Strait of Çanakkale, 258
suffragette, 35
Sunday Girls, 25
Sundsvall, 14, 190, 191, 230, 231,
 252
Superior, full-rigged ship, 26
Suppé, Franz von, 72
Suquamish, 103n82
suspended animation, 30
Svartvik, 190, 230
Svedberg, Carl Israel, 190
Swedish Arrack Punch, 31, 50, 54,
 198, 199, 236
Swedish Vice Consulate in Elsinore,
 253
Sydney, 64n25, 112n87, 199, 259

T

Table Bay, 233
Tacoma, 74n53, 221
Tahiti, 86, 87, 120
Tasmania, 74n51, 121, 141
Taylor, Thomas Rawson, 157n111
Tegnér, Esaias, 155n106
Terserus, Hulda, 18
Thomasson, Olof, 245
Thor, full-rigged ship, 2, 24–26,
 186, 240, 243
Thor, the god of thunder, 180
Thuringia, 125
Tilkhurst, clipper ship, 199
Timor, 85
Tonga, 149
Topelius, Zacharias, 134

Torekov, 254
Torrens, clipper ship, 200
tournure, 150
Trade Wind, 46, 57, 82, 89, 141, 142, 165, 167, 173, 225
Trapp, Axel, 230, 231
Troedson Nordgren, Janne, 241
Tropic of Cancer, 47–49, 166, 204
Tropic of Capricorn, 204
Tropics, 52, 89, 116, 236
Tunis, 235
Turkey (Türkiye), 236, 258
Tuvalu, 141
Twana, 74n53
Twenty Thousand Leagues, 169
typhoid fever, 107, 140, 236

U

Ulleråker mental hospital, 267
Ulysses, 174
Union of Swedish Musicians, 271
United Kingdom, 86, 144, 167, 190
United Kingdoms of Sweden and Norway, 40, 71n47, 235, 241
United States, 96, 99, 220–222, 239, 280
Upland, 31
Uranie, corvette, 4
Ursa Major, 90
Uruguay, 57, 204

V

Valbo, 14
Valparaíso, Chile, 144, 154, 158n112, 170
Vanadis Expedition, 110, 111, 200

Vancouver Island, 221
Verne, Jules, 169
Vestnik (Вестник, of 1880), Imperial Russian man-of-war, 78n60, 79n61
Videnstam, Adolf, 266–268
Videnstam, Greta Elisabeth, 5, 227, 239, 265, 266–270, 274
Videnstam, Kristina Margareta, 266
Vinland, 216n5
Virgin Islands, 44n6
Vladivostok, 79n61
von Post, Hedvig Katarina, 42, 50, 64, 99
von Schéele, Georg, 121, 122, 143, 191n7

W

Wagner-Jauregg, Julius, 262
Waitaha (Canterbury), 207
Waldenström, Paul Peter, 47n9
Walk of Fame, 37
Waltner, Johan Axel, 63, 194, 204, 273
Wanapum, 100n79
Washington, The State of, USA, 74n53, 96, 100n79, 107, 214, 221
Waxin, Anna, née Hallengren, 227, 268
Waxin, Erik Axel, 268
Waxin, Per Bernhard, 191
Webley & Son, 6
Werner, Doris Rosina, m. Broberg, 102n81
Werner, Maria, 64, 102
Western Cape province, 233, 236

Westervik (Västervik), 140, 213, 231, 233, 236–238, 257, 261, 264, 265
West Indies, 29, 166
Westling, O. F., 217n6
Weymouth, 161
whales (cetaceans), 56, 60, 96, 98, 107, 149
White, Smith & Perry, 128
Wifsta Wharf, 190
Williamstown, 123
Williwaw, 159
Winroth, Amalia Mathilda, née Löfström, 10, 11, 15–22, 26–29, 185, 273
Winroth, Elisabeth Charlotta, née Leufvenius, 15, 17–19, 28–30, 41, 42, 48, 50, 62n34, 65, 68, 72, 97, 98, 118, 157, 160
Winroth, Olof, 15, 17, 19n2, 29
Wismar, 231
Wohlfahrt, Bertha and Amelie (sisters), 82, 118, 122, 123
Wohlfahrt girls, 83
Workhouse Street (Arbetarhusgatan), 13
Wright, Karen, 207

Y

Yarra River, 75, 120
Youngberg, Ray, sheriff, 257
YWCA, 35

Z

Zazarenny, V., 79n61
Zealand, 41–44
Zellinger, Catharina Charlotta, 18, 22, 26, 47n10, 68n34, 73n50, 97n74, 232, 235
Zellinger, Evelina, 22
Zellinger, Ida, 22, 41
Zellinger, Knut Reinhold, 22, 36, 47n10, 232, 235
Zellinger, N. Gustaf (Gullich), 13
Zellinger, Pehr, skipper, 13, 18, 47n10, 68n34, 97n94,
Zellinger, Pehr Johan, seaman, 22, 24, 25, 26, 47n10, 215n4
Zellinger's farm, 18
Zetterström's *Gefle-punsch*, 54n16
Zickerman, Ida Cornelia Fredrika, née Sjölander Stenbeck, 71, 73, 82, 83, 118

About the Author

Anders Hallengren is a Swedish essayist and researcher, born in 1950, and the author of some twenty books on historical, international, and philosophical issues. A Harvard Alumnus and Harvard Club member, he received his PhD for a dissertation on Ralph Waldo Emerson's philosophy of nature (*The Code of Concord: Emerson's Search for Universal Laws*, published in 1994) and is an Associate Professor of Comparative Literature at Stockholm University. He is the editor of the widely read WSPC publication *Nobel Laureates in Search of Identity and Integrity: Voices of Different Cultures*, which first appeared in print in 2004. Apart from Europe and the Americas, he has been working in Africa, Asia, and Oceania as observer, lecturer, interpreter, teacher, musician, and news agency reporter. A descendant of seamen of both paternal and maternal lineage he says he feels on safe ground only offshore and at home on tour. To be able to write this book, which has been in progress since 1979, he travelled round the world in the wake of the *Atlantic*. As a writer, scientist, and a scholar, he is a Fellow of the Linnean Societies of London and New South Wales, Australia, and a member of the Swedish Society for Maritime History and the Authors Guild of America. He has been a Visiting Fellow in the Department of History, Harvard University, a Visiting Professor at the University of Hawaii, and for many years served as a founding member, contact, and adviser of the American-Russian Transnational Institute in Norwich, Vermont, and Moscow, along with participating in educational projects in the People's Republic of China. He is a member of the United Nations Association and the Red Cross and received his education in human rights at the OHCHR in Geneva. Apart from being a former volunteer and rescue worker, he has published works on African and Caribbean affairs, and International Law. The theme of his most recent book was the triangular slave trade by sea and human trafficking in our modern cities, showing that the efforts to abolish slavery and slave trade in reality was a complete failure (*The Moment Is Now*, 2019).

www.ingramcontent.com/pod-product-compliance
Lightning Source LLC
Chambersburg PA
CBHW070307230426
43664CB00015B/2667